Patrick Moore's
Practical Astronc

Springer
London
Berlin
Heidelberg
New York
Hong Kong
Milan
Paris
Tokyo

Other titles in this series

Telescopes and Techniques
Chris Kitchin
The Art and Science of CCD Astronomy
David Ratledge (Ed.)
The Observer's Year
Patrick Moore
Seeing Stars
Chris Kitchin and Robert W. Forrest
Photo-guide to the Constellations
Chris Kitchin
The Sun in Eclipse
Michael Maunder and Patrick Moore
Software and Data for Practical Astronomers
David Ratledge
Amateur Telescope Making
Stephen F. Tonkin
Observing Meteors, Comets, Supernovae
and other Transient Phenomena
Neil Bone
Astronomical Equipment for Amateurs
Martin Mobberley
Transit: When Planets Cross the Sun
Michael Maunder and Patrick Moore
Practical Astrophotography
Jeffrey R. Charles
Observing the Moon
Peter T. Wlasuk
Deep-Sky Observing
Steven R. Coe
AstroFAQs
Stephen F. Tonkin
The Deep-Sky Observer's Year
Grant Privett and Paul Parsons
Field Guide to the Deep Sky Objects
Mike Inglis
Choosing and Using a Schmidt–Cassegrain Telescope
Rod Mollise
Astronomy with Small Telescopes
Stephen F. Tonkin (Ed.)
Solar Observing Techniques
Chris Kitchin
Observing the Planets
Peter T. Wlasuk
Light Pollution
Bob Mizon
Using the Meade ETX
Mike Weasner
Practical Amateur Spectroscopy
Stephen F. Tonkin (Ed.)
More Small Astronomical Observatories
Patrick Moore (Ed.)
Observer's Guide to Stellar Evolution
Mike Inglis
How to Observe the Sun Safely
Lee Macdonald
The Practical Astronomer's Deep-sky Companion
Jess K. Gilmour

Nick James and Gerald North

With 105 Figures

British Library Cataloguing in Publication Data
James, Nick
Observing comets. – (Patrick Moore's practical astronomy series)
1. Comets – Observers' manuals 2. Comets – Amateurs' manuals
I. Title II. North, Gerald
523.6
ISBN 1852335572

Library of Congress Cataloging-in-Publication Data
A catalog record for this book is available from the Library of Congress

Patrick Moore's Practical Astronomy Series ISSN 1617-7185
ISBN 1-85233-557-2 Springer-Verlag London Berlin Heidelberg
a member of BertelsmannSpringer Science+Business Media GmbH
http://www.springer.co.uk

Printed in Great Britain

Typeset by EXPO Holdings, Malaysia
Printed and bound at the Cromwell Press, Trowbridge, Wiltshire
58/3830-543210 Printed on acid-free paper SPIN 10844359

Foreword

There has been a recent upsurge of interest in comets in both the amateur and professional arenas of astronomy. Undoubtedly the recent apparitions of two great comets, Hyakutake (early in 1996) and Hale–Bopp (early in 1997), has had a lot to do with this. The wide press coverage given to the collision of comet Shoemaker–Levy 9 with Jupiter in 1994 has also brought to the notice of many people the, albeit remote, possibility of cometary fragments slamming into our own planet. This potential hazard has been highlighted in many television programmes and published articles in the last few years.

This is a very good time to specialise in the observation of comets. However, it is also true that the last few years have seen some fundamental changes in the equipment available to amateur astronomers and a consequent revolution in the types of work that amateurs undertake. Take a look at an observing manual that is just a decade old and you will see how dated it appears.

This book is intended to serve as a guide for the aspiring, modern, comet observer. It is a "primer", especially aimed at those who have some knowledge and experience and who wish to go further, perhaps even to contribute observations of real scientific value.

"Modern" does not always have to mean using expensive equipment to carry out highly technical tasks and some space early in this book is devoted to what can usefully be done with a "low-tech" approach and relatively inexpensive equipment. However, the main bulk of this book does reflect the activities of, and equipment used by, today's advanced amateur astronomers.

We hope that you enjoy this book and find it useful. We especially hope that you will find satisfaction and enjoyment from using whatever equipment you can muster to observe and record the eerie, beautiful and mysterious comets that often arrive in our skies.

Nick James and **Gerald North**

Acknowledgements

The authors would like to offer their thanks and appreciation to the following people and institutions who very kindly allowed us to use their photographs and images in this book and others who took the time and trouble to help us obtain materials: Martin Mobberley, Robert Bullen, Dr Richard McKim, Brian Carter, Case Rijsdijk, Dr H.U. Keller, Jonathan Shanklin, Michael Hendrie, Denis Buczynski, Glyn Marsh, Stewart Moore, Ron Arbour, Steve Goldsmith, Herman Mikuz, Giovanni Sostero, Terry Platt, Lennart Dahlmark, Richard Fleet, Christian Buil, Wil Milan, P. Blasich, Richard Hook, Maurice Gavin, Mauro Zanotta, Meade Instruments Corporation, Orion Telescopes and Binoculars, the British Astronomical Association (BAA), the South African Astronomical Observatory (SAAO), the European Southern Observatory (ESO), the European Space Agency (ESA), The Jet Propulsion Laboratory (JPL), Caltech, Starlink, the SOHO/LASCO Consortium, the Max-Planck-Institut für Aeronomie (MPAE), Project PLUTO, the National Aeronautics and Space Administration (NASA), the Space Telescope Science Institute (STScI).

We are also grateful to have been allowed to feature examples of the work of the late Dr R. Waterfield, George Alcock, and Harold Ridley in this book.

To all of the foregoing we offer our grateful thanks.

Nick James and **Gerald North**

Contents

Chapter 1
Celestial Phantoms

In past times the apparition of a great comet "invading" the hitherto serene heavens struck fear into the hearts of many people. Their fears were born of ignorance and were calmed as science advanced and it was realised that comets were mostly very insubstantial things. Recently, however, the tide has to a small extent turned back. This time it is the scholars who are the most worried. Lately, there has been much publicity concerning the possibility of a large asteroid hitting the Earth and the consequences that would ensue. Less well publicised is the possibility that the solid part of a comet might do the same thing. It **may** even be the case that some or all of the mass extinctions that depopulated our planet in the past tens and hundreds of millions of years happened because of comets! If so, it is likely to happen again sometime in the next few million years.

Comets also tell us something about the formation of our Solar System and the conditions in it when it was newly born. They are also fascinating things in their own right. Their appearances can range from small, faint, grey, fuzzy patches even when seen through large telescopes, to large filamentary forms tinted with pale orange, yellow, green and blue and spanning much of the sky, easily visible to the unaided eyes of the admiring observer. Those of the latter type are rare but they do happen. You probably remember the comets of the spring of 1996 and 1997. Most are not that spectacular but there are plenty more comets which are bright and large enough to be impressive when seen with any form of optical aid. Their observation and study is both a fascinating and a scientifically valuable exercise.

Ancient Apparitions

We certainly do not know who was the first human being ever to see a comet but one thing we can be sure of is that comets have been recognised as occasional visitors in the sky for thousands of years. The ancient Chinese certainly saw and recorded what they called "broom stars" and in fact our modern term "comet" is derived from the Greek "aster kometes", meaning "long-haired star". Comets come in different sizes, vastly different brightnesses and, to a certain extent, different shapes. In general they appear as filmy or misty "stains" of greyish light against the sky, with one end of the apparition, the *head* being the brightest and most compact part. Often the head contains a bright, star-like, speck of light. The *tail* often fans out from the head, or sometimes extends out as a narrow trail of milky gossamer.

Figures 1.1 to 1.9 shows some comets as they were recorded during the twentieth century. Notice the

Figure 1.1. Comet C/1956 R1 Arend-Roland photographed by R.L. Waterfield on 1957 April $13^{d}\ 24^{h}.9$ UT, using an f/4.3 Cooke triplet lens of 152mm aperture and 0.66m focal length with an exposure of 35 minutes on a Kodak Oa-O photographic plate.

Figure 1.2. Comet C/1959 Y1 Burnham photographed by M.J. Hendrie and H.B. Ridley on 1960 April $27^d.1$ using R.L. Waterfield's equipment as detailed for Figure 1.1. This time the exposure was 30 minutes on a Kodak Oa-O photographic plate.

variety of forms they displayed. You will find a much larger selection of images in the CD-ROM that accompanies this book. A comet's form is not constant but rather changes and evolves as it moves into our skies and reaches it greatest prominence, however great that turns out to be. Later in this book we will more fully discuss comets ranging through all the possible forms and brightnesses.

As one might expect, the comets of long ago were regarded as things astrological, rather than astronomical. They were taken to be signs of coming momentous events, perhaps warnings from the gods, and most often as portents of doom and tragedy. Even today we still call the visitations of comets in our skies *apparitions*. The brightest comets caused widespread panic among people. This was probably exacerbated by great comets making appearances at times like 1066, on the eve of the Battle of Hastings, at 1665 as the Great Plague was

Figure 1.3. Comet C/1969 Y1 Bennet photographed by M.J. Hendrie on 1970 April $4^{d}.2$ using R.L. Waterfield's equipment as detailed for Figure 1.1 for a 40 minute exposure on a Kodak Oa-O photographic plate.

taking hold in England, and again in 1666 just before the Great Fire that was to purge the plague-ridden London!

Amazingly, the nonsense of astrology still lingers strong in the minds of many people, as does the belief that comets represent an aspect of the supernatural. Harmless enough, you might say. Ordinarily this is so. However, one terrible tragedy of recent times occurred during the apparition of Comet Hale–Bopp. The members of a religious sect committed mass suicide because their "spiritual leader" had told them that the comet heralded spiritual beings from outer space riding in a flying saucer alongside the tail of the comet. Apparently they all killed themselves in order that their souls could join the extraterrestrials.

While mysticism pervaded and persisted in the minds of many people down the centuries, there were at least some – the scientists and the great thinkers – who were also trying hard to figure out what comets

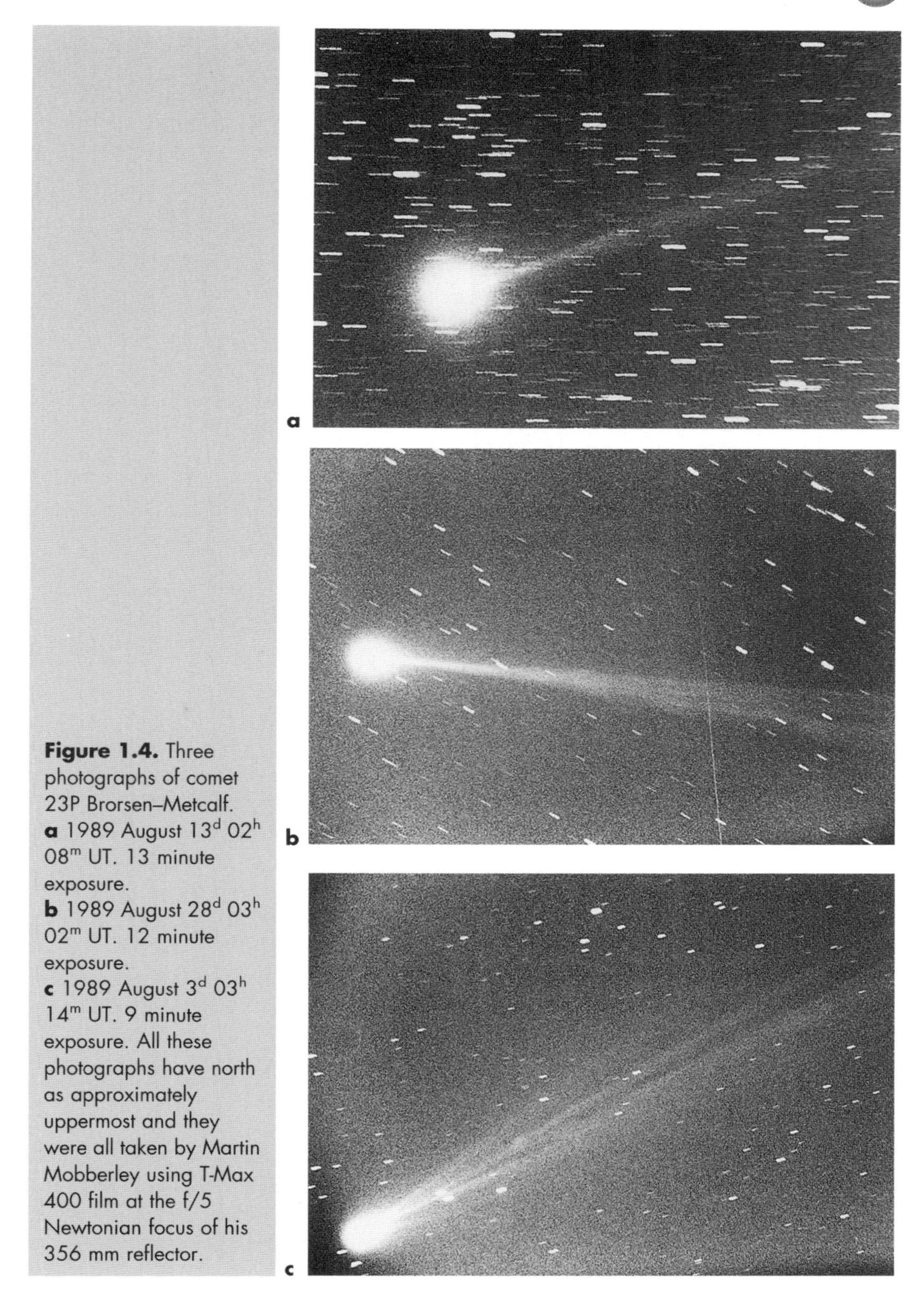

Figure 1.4. Three photographs of comet 23P Brorsen–Metcalf.
a 1989 August 13^d 02^h 08^m UT. 13 minute exposure.
b 1989 August 28^d 03^h 02^m UT. 12 minute exposure.
c 1989 August 3^d 03^h 14^m UT. 9 minute exposure. All these photographs have north as approximately uppermost and they were all taken by Martin Mobberley using T-Max 400 film at the f/5 Newtonian focus of his 356 mm reflector.

really are and where they come from. About three-and-a-half centuries BC, the great philosopher Aristotle had proposed that comets are objects ejected from the Earth

a

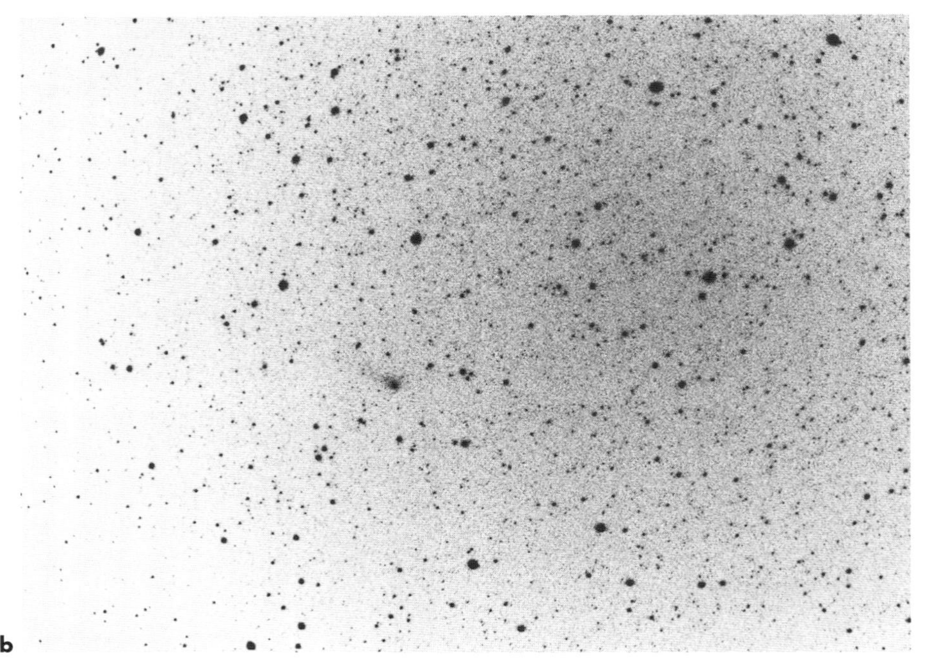

b

Figure 1.5. Comet C/1987 P1 Bradfield. **a** Taken by Martin Mobberley on 1987 November $14^d\ 18^h\ 03^m$ UT. The 17 minute exposure was made on Tri-X film at the f/5 Newtonian focus of his 356 mm reflector. **b** The comet was well on the wane for this photograph taken on 1988 January $14^d\ 18^h\ 45^m$ UT by Gerald North. This 10 minute exposure was made on 3M Colourslide 1000 film, using an ordinary f/2.8 135 mm telephoto lens. The camera was guided at the stellar rate by mounting piggyback onto a driven telescope. The photograph is here reproduced in negative. This is a technique often used in astronomy as ''faint fuzzies'' are easier to detect when appearing as dark against a light background.

Figure 1.6. Comet C/1989 X1 Austin. 1990 April 29^{d} 02^{h} 05^{m} UT. 13 minute exposure commencing at that time on T-Max 400 film at the f/5 Newtonian focus of a 356 mm reflector. Martin Mobberley.

and are swept along by the motion of the sky, catching fire in the process. Aristotle's opinions about things held great sway at the time and for centuries after his death. As absurd as the idea now seems to us, one can argue that it was not until after Tycho Brahe's practical investigation in 1557 that Aristotle's idea about comets was finally laid to rest.

Tycho Brahe was a most peculiar individual. A Dane of noble birth, he was eccentric and quarrelsome. One of his many altercations with people resulted in the lower part of his nose being badly disfigured. From then on he wore a home-made prosthetic made of silver, gold and sealing wax! He set up a private observatory (funded by the King of Denmark), called "Uraniborg", on the island of Hven (close to the mainland and not far from Copenhagen), where it is said he kept a dwarf in the manner of a domestic pet. The King had installed him as landlord on the island, which was rather unfortunate for his tenants as Tycho proved to be a tyrannical, even cruel, overlord.

Still, it is for his contributions to astronomy that we most remember Tycho today. At Uraniborg he made measurements of the positions of celestial bodies with self-designed apparatus. His results were of unprece-

Figure 1.7. Comet C/1990 K1 Levy appearing close to the globular cluster M5 on 1990 August 18^d 21^h 30^m UT. 8 minute exposure commencing at that time on T-Max 400 film. 356 mm f/5 reflector. Martin Mobberley.

dented accuracy. This is especially amazing because his determinations of positions were, of necessity, made by sighting his instruments by eye. There were no telescopes or other optical instruments he could use to aid him.

Tycho measured the positions of 777 stars, achieving measures accurate to better than 4 minutes of arc. His practical researches touched on many areas of positional astronomy. He followed the moving planets in their paths across the heavens. Johannes Kepler, who worked with him in Tycho's last years, made use of the copious results from Uraniborg in studying the motions of the planets. Kepler eventually developed his famous Laws of Planetary Motion as a result of the database of measurements. In turn, it was partly as a result of Kepler's Laws, that Isaac Newton was able to develop his Universal Law of Gravitation.

Returning to Tycho's investigation of comets, he reasoned that if comets are phenomena occurring in the Earth's atmosphere they should show the effects of parallax. That is, if a comet is observed over several

Figure 1.8. Comet 109P/Swift–Tuttle. 1992 November 13^{d} 18^{h} 16^{m} UT. 24 minute exposure commencing at that time on hypersensitized Kodak TP2415 film. 356 mm f/5 reflector. Martin Mobberley.

hours it should shift its position against the background pattern of stars due to its proximity (as the celestial bodies move in the diurnal paths across the sky the nearby comet would be seen in different directions against the more distant star background). Of course, a comet also has its own real motion across the sky, so the two effects have to be disentangled but this can be done.

Altogether, seven comets were visible during Tycho's time at Uraniborg. He made careful measurements but found no parallax, proving to his satisfaction that the comets exist well beyond the Earth's atmosphere. He concluded that comets were at least as far away as the Moon, and probably a lot further. Kepler championed Tycho's ideas, even writing a book about comets. Even so, it took a while for Tycho's ideas about comets to be accepted but the truth eventually did filter through.

The next major advance came when Edmond Halley, the noted scientist and England's second Astronomer Royal, used mathematical methods, based on Newton's Law of Gravitation, to plot the paths of a number of great comets. In fact, he collected observations on 24 comets but the major breakthrough came as a result of just one of them. He determined that the impressive comets of 1531, 1602 and 1682 travelled along the same path. In fact, he realised that they were one and the

a

same object. This comet whirled round the Solar System in an elliptical path, taking 76 years to do one "round trip" (see Figure 1.10).

Halley made the prediction that the comet would return in 1758. It did – being recovered by Johann Palitzsch on Christmas night in that year. It is a shame that Halley had died 14 years before his vindication because his breakthrough had, at a stroke, told us something about the nature of comets and their true behaviour, as well as supplying independent verification of Newton's Universal Law of Gravitation and its application to the science of celestial mechanics.

Thanks to his momentous discovery, the great comet of 1682 (and 1531, 1602 and 1758) became known as Halley's Comet. This is certainly fitting. More usually, though, a comet is given the name of the first person to see it. This has been the scheme used from the earliest

Figure 1.9. Two photographs of comet C/1996 B2 Hyakutake by Gerald North, taken with an ordinary camera, fitted with a 58 mm f/2 standard lens. **a** A deliberately framed "pretty" shot. The comet's proximity to the celestial pole at the time (1996 March 27^{d} 21^{h} 10^{m} UT) allowed this 1 minute exposure to be given with the camera mounted on a simple undriven tripod.

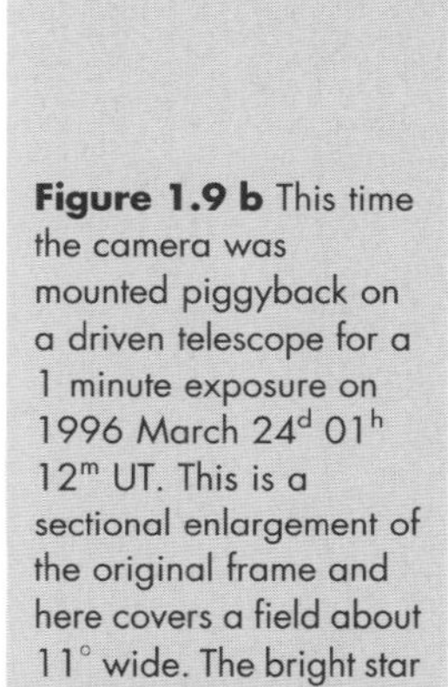

Figure 1.9 b This time the camera was mounted piggyback on a driven telescope for a 1 minute exposure on 1996 March 24^d 01^h 12^m UT. This is a sectional enlargement of the original frame and here covers a field about 11° wide. The bright star near the head of the comet is γ Boötes

b

days right to the present time. There are complications to this scheme, and there are other schemes used today which you should also know about but we can defer such matters until the end of this chapter. For now we can continue our introductory survey of the visiting celestial phantoms that are the comets.

Comet Orbits

All bodies, including comets, which orbit the Sun obey Kepler's Laws of Planetary motion:

1. The bodies move in elliptical orbits, with the Sun located at one of the foci.
2. The radius vector (line joining the Sun to the body) sweeps out equal areas in equal times.
3. The squares of the sidereal periods (orbital periods) of the bodies are proportional to the cubes of the semi-major axes of their orbits (the semi-major axis = half the diameter of the ellipse measured along its greatest length).

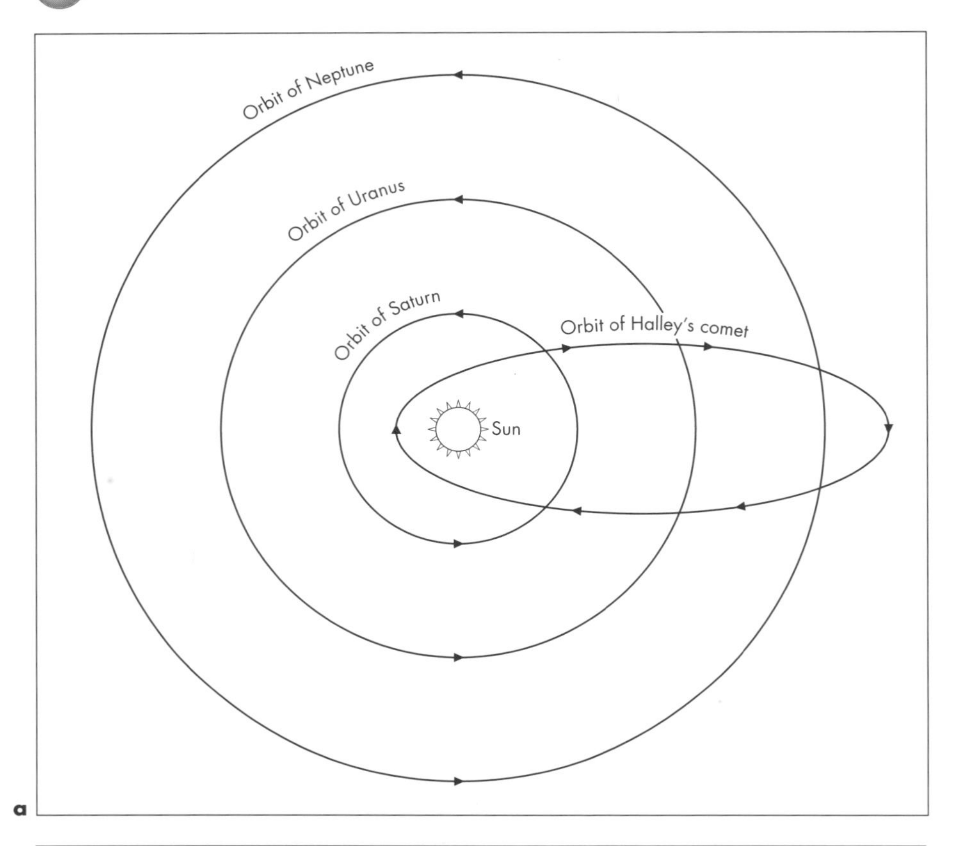

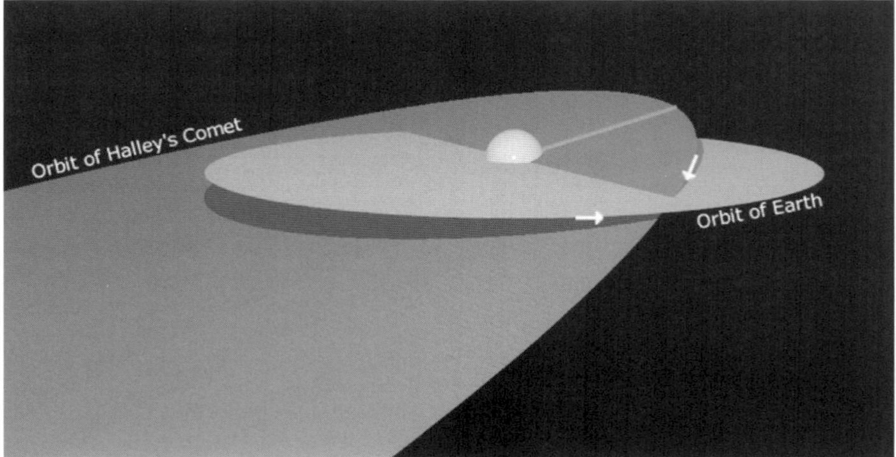

Figure 1.10. a A plan-view of the orbit of Halley's Comet. **b** A 3-dimensional representation of the orbit of Halley's Comet showing its inclination to the orbital plane of the orbit of the Earth (which is also very similar to the planes of the other major planets out to Neptune).

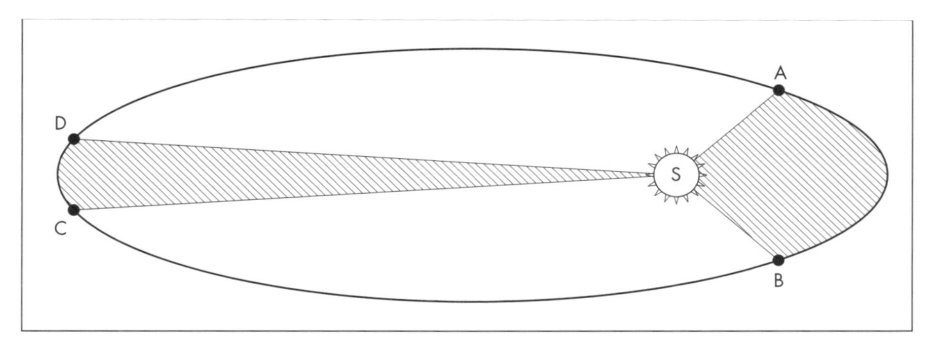

Figure 1.11. Kepler's Laws. The comet moves from A to B in the same time as it moves from C to D (see text for details).

Basically, the further out the body is from the Sun, the slower it moves. The planets move in paths which have relatively low eccentricities; that is to say their orbits closely resemble circles, even though strictly speaking they are ellipses. By contrast, most comets move in highly eccentric orbits.

Imagining that a comet moves along the path such as that illustrated in Figure 1.11, the comet varies its speed as it goes around its orbit so that the area swept out as the comet moves from A to B is the same as that swept out as it moves from C to D. Put another way, the comet moves very slowly when it is far away from the Sun but its speed increases as it nears the Sun. The comet's speed is greatest when it is nearest the Sun, or as we say at *perihelion*. The comet slows down as it recedes from the Sun and it is moving at its slowest when it is farthest from the Sun, or *aphelion*.

The result is that a comet spends by far the greatest amount of time well away from the Sun and only a very short time near to it while it is passing through perihelion.

Referring again to Figure 1.10, you might notice that Halley's Comet moves in the opposite sense to that of the main planets shown on the diagram. Halley's comet may have been the first to have the shape of its orbit properly defined but it was certainly not the last. It turns out that about half of all comets orbit the Sun in the "wrong way" or retrograde sense. The planets all orbit in roughly the same plane about the Sun but comets have orbits which can be inclined at any angle to the plane of the planetary motions.

Something else was revealed when enough comets had their orbits determined: comets seem to be divided into two main classes with regard to their orbits. Some comets frequently return to perihelion, taking just a few years or a few tens of years. These we call *periodic comets*. Others take many centuries, or thousands of years, and even longer, to return to the inner realms of the Solar System. These are known as *long-period comets*. All the long-period comets have exceedingly eccentric orbits whose aphelia lie at enormous distances from the Sun. The Earth–Sun distance is 150 million km, also known as 1 Astronomical Unit (AU). The aphelia of long-period comets often lie at distances of the order of 50 000 AU from the Sun. Most of the comets we detect are of the long-period variety.

In order to fully appreciate a subject one should know something about its roots and it is in that spirit that we have offered this highly abridged, whistle-stop, tour of the beginnings of the study of comets. The resources section at the end of this book (in the Appendix) lists a number of works concerned with the history of comet observing and research that you might also like to consult.

Before we go any further, it is now time to deal with the slightly complicated matter of how comets are identified by means of names and numbers ascribed to them.

Cometary Nomenclature

In most cases a comet is given the name of the first person to see it. This has been the scheme used from the earliest days right to the present time. So, for instance, the great comet discovered by Jack Bennett in 1969 became known as Comet Bennett. In these cases of a single discoverer a slightly different style is sometimes used: in this example it would be "Bennett's Comet". However, in scientific literature it is the first scheme which is preferred.

If more than one person discovers the comet independently (the International Astronomical Union are the arbiters in all cases) then the comet will carry the names of the co-discoverers, arranged in their order of discovery. There is one caveat: comets discovered by means of space probes or those by automated search programmes bear the name of the probe, rather than the human being sitting at a console in front of a screen who made the discovery. Such was the case with Comet *IRAS*–Araki–Alcock which became notable for its relatively close passage past the Earth in May 1983. It was first detected in the data downloaded from the *Infrared Astronomical Satellite* (*IRAS*), then by the Japanese

astronomer Genichi Araki, and then by the English veteran comet and novae discover George Alcock.

Since some comet hunters discover more than one comet in their lifetimes, it is not enough just to identify a particular comet by the discoverer's name. More detail is needed.

Prior to 1995 the scheme in general use was to give a particular comet an additional designation comprising its year of discovery and the order of discovery in that year. For instance, the thirteenth comet discovered by William Bradfield was designated 1987s, because s is the nineteenth letter in the alphabet and his comet was the nineteenth to be discovered in 1987.

In due time, when enough observations of a newly-observed comet had been made so allowing its orbit to be firmly established, it was given yet another designation, comprising the year of its discovery and a number denoting the order of perihelion passage that year. Yet another piece of information was added if the comet was of the periodic class: the letter P/ prefixed the moniker. Hence the famous short-period "Encke's Comet" would be properly designated as: P/Encke 1786 I, because it was the first of the comets discovered in 1786 to pass perihelion. The IAU changed the scheme for designating asteroids and comets in January 1995. Now we must always use the modern scheme. However, 1995 is not that long ago and so you may well have to deal with the older scheme in books and other literature, which is why I have devoted some space to it here.

The Modern (IAU Approved) Scheme for Cometary Nomenclature

First we have a letter "C", or a letter "P", then followed by a forward slash. "P" is used if the comet is "periodic", that is: determined to have a period of less than 200 years. "C" is used for all comets of longer period. There are two other letters that could be used in place of the "P" or the "C", though those instances are quite rare. "D" is used for comets that have disappeared, or no longer exist and "X" is used for those who orbits cannot be computed owing to a lack of observations.

Next comes the year of discovery and then a space followed by a letter. The letter denotes the half-month of the discovery according to the following:

A	January 1–15	B	January 16–31
C	February 1–15	D	February 16–29
E	March 1–15	F	March 16–31
G	April 1–15	H	April 16–30
J	May 1–15	K	May 16–31
L	June 1–15	M	June 16–30
N	July 1–15	O	July 16–31
P	August 1–15	Q	August 16–31
R	September 1–15	S	September 16–30
T	October 1–15	U	October 16–31
V	November 1–15	W	November 16–30
X	December 1–15	Y	December 16–31

The letter "I" is not used in case it is confused with the number 1 and, as you can see, the letter "Z" is not needed.

The next component of the designation is a number which indicates the order of discovery in the denoted half-month.

Lastly we have the name of the discoverer(s).

Periodic comets do not carry a date of discovery but instead have a prefix number indicating the order of discovery of their periodic status. This starts with Halley's Comet, which becomes 1P/Halley. The second periodic comet is Encke's, so this is properly denoted 2P/Encke.

In case this seems complicated, let me give you an example. Remember the bright comet that graced our skies in the spring of 1996? That one was: C/1996 B2 Hyakutake, meaning that it was the second comet (2) discovered in the second half of January (B) in 1996. The orbit of this comet has a period of more than 200 years (C/) and its discoverer's name was Hyakutake (actually, the Japanese amateur Yugi Hyakutake). Enough said?

Here are the other comets I have mentioned earlier in this chapter properly denoted using the modern IAU approved scheme of nomenclature: C/1995 O1 Hale–Bopp; C/1969 Y1 Bennett; C/1983 H1 IRAS–Araki–Alcock; C/1987 P1 Bradfield.

This book is mainly concerned with the practical matters involved in the discovery, observation and study of comets. Before we finally get down to that, it will benefit us to get a basic overview of our modern understanding of the nature, structure, physics, and chemistry of comets and how they interact with the interplanetary environment. That is the subject of the next chapter.

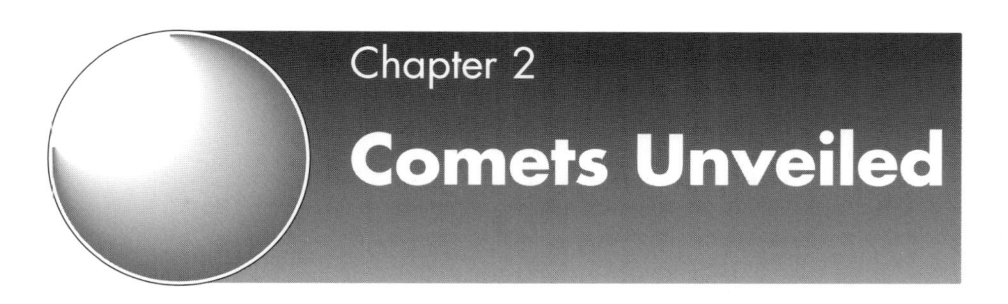

Chapter 2 Comets Unveiled

Comets might have been known for millennia but it is only in the last few decades that we have really got to know some of their most intimate physical and chemical secrets. We had to wait until the mid-1980s before man-made machinery approached comets at close range, sniffing and sensing, and sending back surveillance. Now, at long last, we have a good picture of what a comet is and how comets behave. Even so, there are still secrets to be uncovered and truths to be clarified, especially about how they fit into the grand scheme of things. That is what makes the observation and study of comets so worthwhile.

A Picture Emerges

The invention of the telescope allowed astronomers to augment their probing of the structures of comets beyond what is possible with the naked eye. By the late nineteenth century long-exposure photography was brightly illuminating our previously dim views and allowing us to probe much further into the dynamic structure of comets.

Though the visual and photographic images told us a lot, we needed to apply more sophisticated techniques in order to tease the more intimate details from comets. Spectroscopy, carried out in wavelengths spanning much of the electromagnetic spectrum, has revealed to us most of what we have learned about the detailed chemical and physical natures of comets. Imaging comets at all wavelengths of the electromagnetic spectrum, in addition

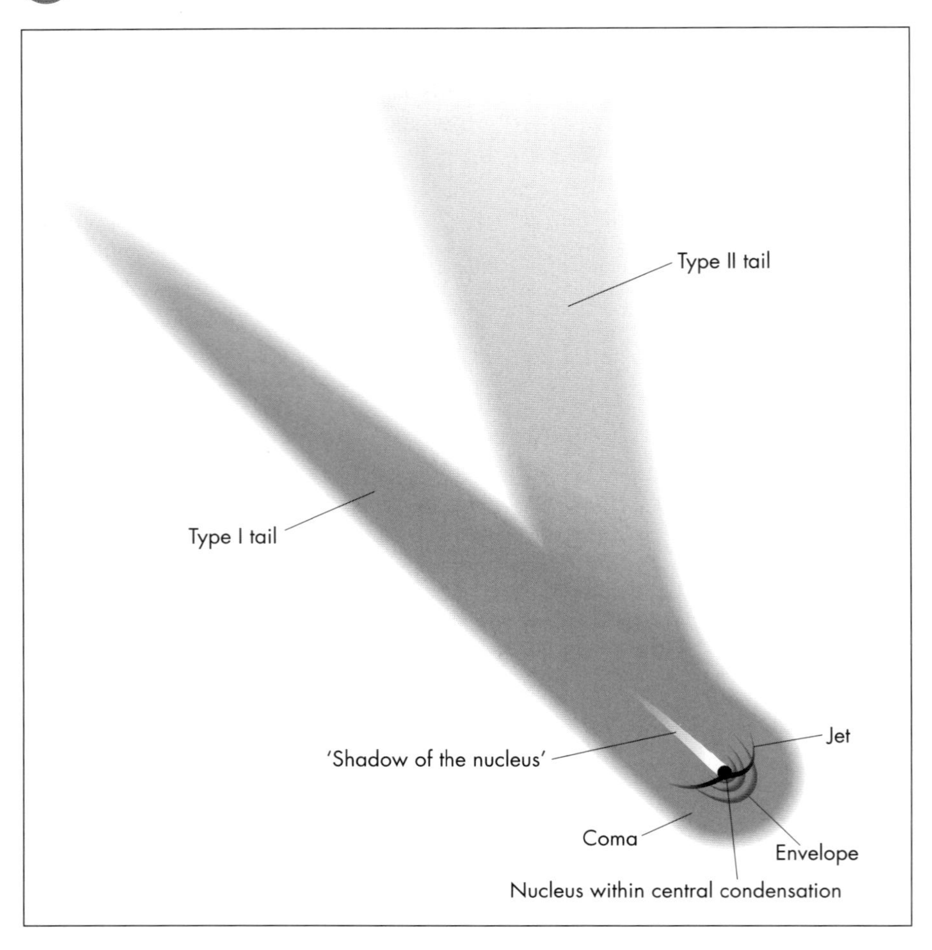

Figure 2.1. Representation of the structure of a hypothetical comet.

to the visual range – sometimes of necessity from above the Earth's atmosphere – has also added much to our knowledge.

Though comets vary greatly in appearance from one to another, and any given comet changes its appearance with time, we can settle upon a simplistic representation of the overall structure of a bright comet as it appears when near perihelion (see Figure 2.1).

In this book about observing practice there is only room enough to give a very sketchy overview of what we know about the physics and chemistry of comets – but observations beyond simple sight-seeing become diminished in their point and purpose when divorced from the science. So, in the following sections we will take just a little space to look at the various features of comets in more detail.

A Comet's Nucleus, Jets and Shells

Astronomers long ago realised that most of a comet is very insubstantial. Indeed, most of what we perceive as a comet we would, in other circumstances, regard as a vacuum! However, there is one part of a comet that, despite being dwarfed in size by the rest of it, is very substantial in nature: the *nucleus*. The vast bulk of all the matter of the comet is concentrated in its nucleus. A large cometary nucleus may have a mass of around several thousand million million kilograms (10^{15} kg in scientific notation).

The nucleus resides in the head of the comet and it consists of what we nowadays refer to as a "dirty snowball" of silicate rock and ices. It is irregularly shaped and is a few kilometres across.

As far as we can tell, the nucleus of a comet is mostly composed of various ices, with water ice forming the major constituent. The other ices are mostly frozen organic volatiles. Perhaps there might be a small rocky core at the very heart of the nucleus but this seems unlikely, as we think that the comets were made well beyond the realm of the major planets in the Solar System. It is only close in to the Sun that substantial amounts of the refractory materials are to be found.

A typical long-period comet spends most of its time well away from the warmth of the inner Solar System – further from the Sun than even the frigid worlds Neptune and Pluto. With an ambient temperature of just a few degrees above absolute zero (absolute zero = –273°C), the comet must then consist of an inert and frozen ball of "dirty" ice.

It is not until the comet heads inwards that the warming rays from the Sun begin to drive the most volatile of the ices to sublime from its surface, so forming the coma and the tails of dust and gases that we actually recognise as a comet.

In case you are wondering whether or not the ices on the nucleus can melt to a liquid, the answer is no. In any environment of extremely low pressure, ices pass straight from their solid phase to their vapour (gaseous) phase. We say that the ices *sublime*. You might recall that iodine and ammonium chloride are two common chemicals that sublime even at Earth-normal atmospheric pressure.

The gas release tends to be localised, mostly emanating from the surface of the nucleus in sprays of matter that we call *jets*. We know this because many bright comets display jets that can be seen through even the lowliest forms of optical aid (see Figure 2.2). Thanks to photography, jets can be studied in many more comets so we know that they are a common feature.

Comet nuclei are irregular in shape. They also rotate. Both effects can produce brightness changes in the nucleus, though getting a value of the rotation period from brightness variations is not at all straightforward. This is especially so because the nucleus is imbedded in the very much brighter "false nucleus" – and this is dependant of the activity of the comet (which is itself variable). Fortunately, the presence of jets helps us determine the rotation period since they emanate from virtually fixed patches on the surface of the nucleus. Observations indicate that typical comet nuclei have rotation periods in the range of a few hours to a few days.

If one did not suspect that it is the action of the Sun's radiation that drives the jets, then evidence for this is provided by careful study of them as the nucleus rotates. The jets vary their activity, being most vigorous when they are rotated onto the sunward side of the nucleus. The "lawn sprinkler" effect of the jets from a comet's rotating nucleus and the variations in the rate

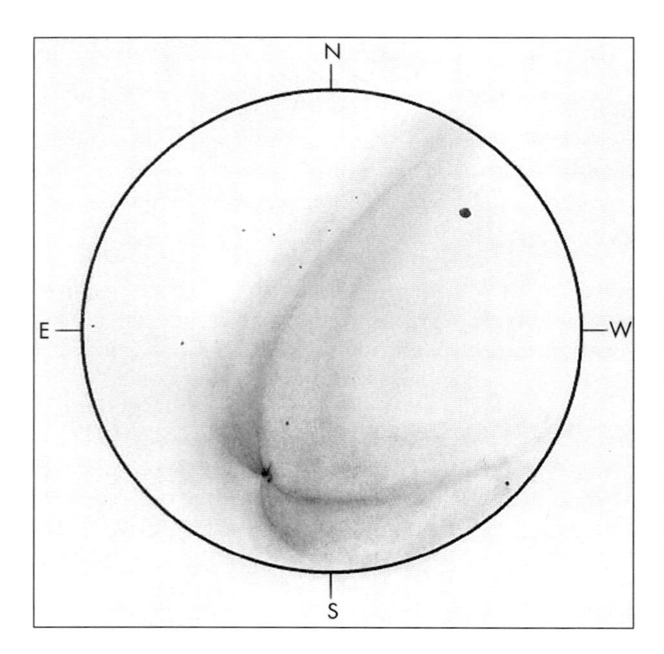

Figure 2.2. Jets from the nucleus of C/1995 O1 Hale–Bopp, as drawn by Robert Bullen.

Date: 8th February 1997
OBS: 31
Log: 18
GMT: 04:00–06:10
Cond: III
Sky: IV–III (Light mist interfered with OBS when comet low)
RA: 19h 59.8m
Dec: + 19° 12′
8.5-inch reflector, X64
Field size: 38 mins
Mag: 2.3

of materials shed into its coma combine with the compressive "wind-sock" action of the solar wind to produce *shells* within the coma. These are mostly evident on the "up-wind" side of the coma, as is shown in Figure 2.3.

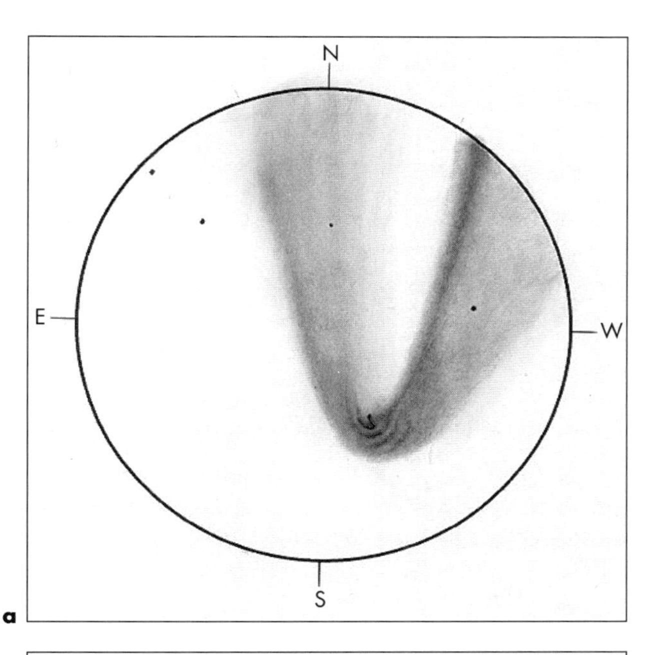

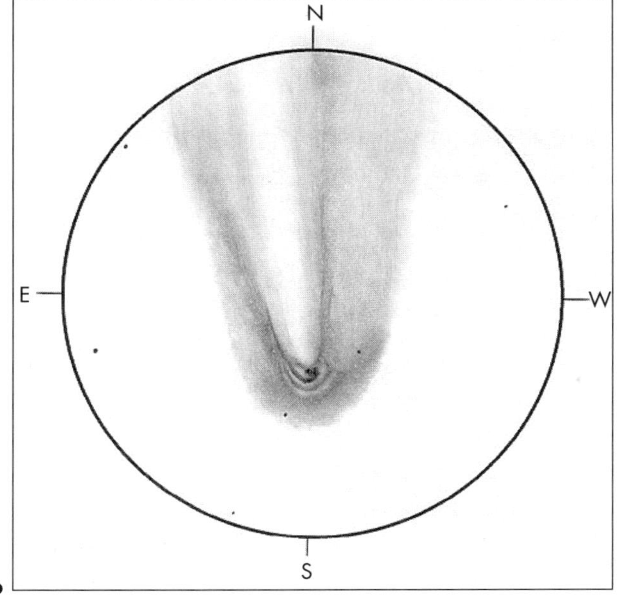

Figure 2.3. Shells in the coma of C/1995 O1 Hale–Bopp as drawn by Robert Bullen on three dates.

a
Date: 28th March 1997
OBS: 53
Log: 33
GMT: 20:00–21:45
Cond: III
Sky: III > IV by 21:15. Some low and high level cloud interfered with the OBS at times
RA: 1h 17.0m
DEC: +45° 23′
6.25-inch reflector, X48
Field size: 50′
Mag: -1.3

b
Date: 2nd April 1997
OBS: 56
Log: 38
GMT: 20:40–22:00
Cond: III
Sky: III with some slight haze
RA: 2h 04.7m
DEC: +43° 45′
6.25-inch reflector, X48
Field size: 50′
Mag: -1.0

(Figure 2.3c is on the following page).

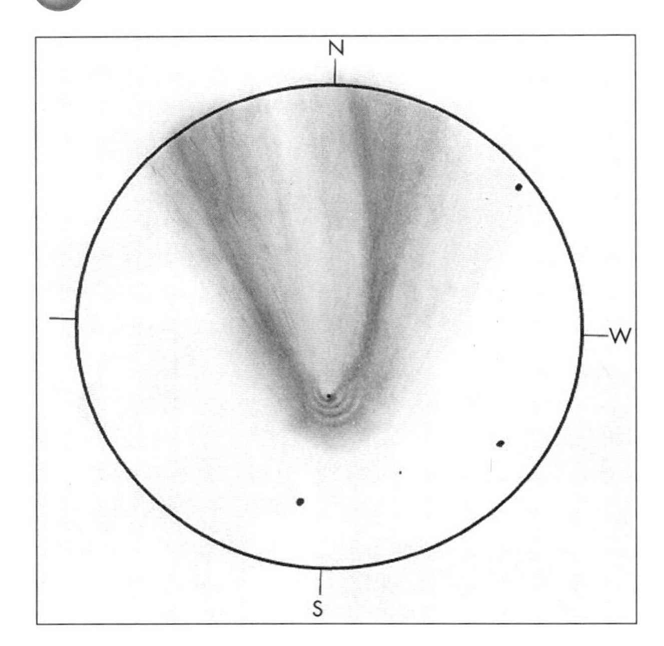

Figure 2.3c.

c
Date: 17th April 1997
OBS: 58
Log: 43
GMT: 20:45–22:30
Cond: III
Sky: III with some light mist and a little interference from light pollution on horizon
RA: 2h 45.0m
DEC: +41° 12′
6.25-inch reflector, X48
Field size: 50′
Mag: -0.1

Interestingly, the presence of jets solves one mystery that had troubled early astronomers: comets often slightly defied Kepler's Laws of Planetary Motion. Astronomers, having gone to much trouble to determine their orbits, were irked that comets refused to stick exactly to the paths mapped out for them! The reason is that these jets, acting like rocket engines, were able to nudge the nucleus about by just enough, over a period of time, to produce a slight change in the orbit of the comet.

A Comet's "False Nucleus" and Coma

The coma of a comet can be thought of as the "atmosphere" surrounding its nucleus. Actually, the outermost part, the *exosphere*, of a planetary atmosphere might best serve for our dubious comparison. The coma of a typical comet is a roughly spheroidal volume of gas and dust, extending to perhaps 100 thousand, maybe exceptionally a million, kilometres from the nucleus. The coma is most tenuous at its outer limits and is most dense near the imbedded nucleus.

In fact, the bright speck that is often seen at the heart of a comet's coma is **not** the nucleus. Instead, we see the greatest concentration of material in the coma. We call this the *false nucleus*. The dim true nucleus is buried inside the brighter false nucleus.

The coma is ephemeral in nature. At distances far from the Sun the nucleus is too cold to have any atmosphere. As it is warmed on approach towards the Sun the most volatile of the chemical ices sublime first, others adding to the chemical smorgasbord as the surface of the nucleus continues to warm up.

Typically, carbon monoxide and carbon dioxide begins to be released from the nucleus in significant amounts when it approaches to within about 10 Astronomical Units (AU) from the Sun. Less volatile ices do not sublime to any large extent until the nucleus gets rather closer to the Sun. For instance, water-ice only begins to sublime significantly when the nucleus approaches within about 3 AU.

Even a massive comet nucleus will have an escape velocity, measured at its surface, of no more than 5 m/s. The gas in the coma is warmed as the comet heads towards the Sun. The average kinetic velocities of the gas atoms in the coma will range into hundreds of metres per second as the comet nears the Sun to within one or two Astronomical Units (AU).

To understand why, I should perhaps explain that the molecules of any gas at a particular temperature and pressure are rushing around randomly in whatever vessel contains the gas. When the molecules strike the walls of the container their combined effect creates what we would measure as the pressure of the gas. In the case of planetary atmospheres, gravity serves the same function, though in a different way, as the walls of a container.

The velocity of a molecule, which is really a statistical average value – known as the "root mean square" value – within a given amount of the gas depends both on the temperature (the velocity increases with increasing temperature) and the mass of the molecule (the velocity is smaller for the more massive molecules). This behaviour forms part of what we call *The Kinetic Theory of Gases* and more about this subject can be found in many advanced ("Year 12") school textbooks on physics, or any general physics text book intended for university undergraduate students.

The particles, comprising the vapours, that enter the space around a comet's nucleus cannot be contained by the extremely weak gravitational field of it and so they stream off into space. The rate of material loss increases with ambient temperature, and therefore with the decreasing distance of the comet from the Sun.

So, the mass and composition of the material in a comet's coma is not at all constant. The following molecules and radicals are, though, usually to be found in the fully-formed comets we observe: water (H_2O); carbon monoxide (CO); carbon dioxide (CO_2); methanal – old name, formaldehyde (HCOH); methyl cyanide – also known as ethanenitrile, old name, acetonitrile (CH_3CN); methane (CH_4); ammonia (NH_3); hydrogen cyanide (HCN); carbonyl sulphide (OCS); various dihydrides such as CH_2 and NH_2; the hydroxyl radical (OH); and other radicals, such as CH and NH; also there are very small amounts of various elements in atomic form such as hydrogen (H), carbon (C), oxygen (O), nitrogen (N) and sulphur (S). If the list seems a long one, it is most certainly not complete – but I have given the major constituents usually to be found in a comet's coma.

However, the list is further extended because the ultraviolet and other short wave radiations from the Sun cause some *ionisation* of the various chemical species. By this I mean that the energetic solar photons can cause electrons (which are negatively charged particles) to be "shaken free" of some of the atoms in certain molecules, leaving them positively charged (they were electrically neutral beforehand). This is not the only mechanism in operation. The solar wind particles colliding with the gases in the coma also cause ionisation. In this way species such as: H_3O^+, H_2O^+, OH^+ – the so-called "water group"; and organics such as: CO_2^+, CO^+, C^+ , CH^+, and many others, including the nitrogen based, so-called "ammonia group", ions are created.

The chemistry is still further complicated because the energy imparted to the molecules or radicals in the comet by the solar photons can go beyond causing them to ionise. Sometimes whole molecules (or radicals) can split into sub-species (this process is called *photo-dissociation*) and so new species are born of the destruction of a small number of the molecules of a "parent" species.

As well as the gaseous species there are also solid grains of matter. These astronomers refer to as *dust*, though the dust in comets is totally unlike the domestic variety. The dust particles are variously composed of various silicates, silicon, carbon, magnesium and sodium, together with some more exotic solid species such as CHON (carbon–hydrogen–oxygen–nitrogen), as well as many others.

The source of these particles must be the nucleus, so how do they find their way into the coma? It is the jets of material spraying off the nucleus that sweep up solid particles from its dusty surface and carry them into the outwardly expanding envelope of gases. You can see that a comet's coma is a dynamic and very complicated system!

The coma of a comet is rendered visible to us mainly because of sunlight scattered by molecular fragments such as C_2, C_3, NH_2, H_2O^+, and CN, as well as by the sunlight reflected and scattered by the solid grains. Hence the colour of a cometary coma tends to be predominantly yellowish-brown.

As stated earlier, the materials comprising the coma are ephemeral. Originating from the nucleus as the comet comes out of its "deep freeze", some of the materials expand radially away, while others are channelled into the tail, or tails.

A Comet's Tails

Yes, **tails** – for most comets sport not one but two tails, each with a different composition. There is even a third type sometimes seen. They may be blended together, or instead well-separated and distinct.

Type I tails are composed of ionised gases, while *type II* tails are made of the fine solid particles which astronomers call dust. Physicists often refer to a gas which is composed of ionised particles as a *plasma*. So type I comet tails are also known as *plasma tails*.

Spectroscopy reveals that type I tails contain ionised molecules and radicals, such as CO^+, CO_2^+, OH^+, CH^+, CN^+, as well as many others. Any chemist will tell you that these radicals are extremely reactive. Under normal Earthly conditions they will aggressively react with any surrounding ions or radicals in an effort to form electrically neutral species. Although some of these exotic species can fleetingly form as intermediates during chemical reactions, we generally never see them on Earth. However, in the cold and near-vacuum environment of a comet's tail it is relatively rare for the radicals to meet suitable partners and so they remain uncombined.

The ionised gases of the type I tails can absorb short-wave radiations from the Sun. They then re-radiate the

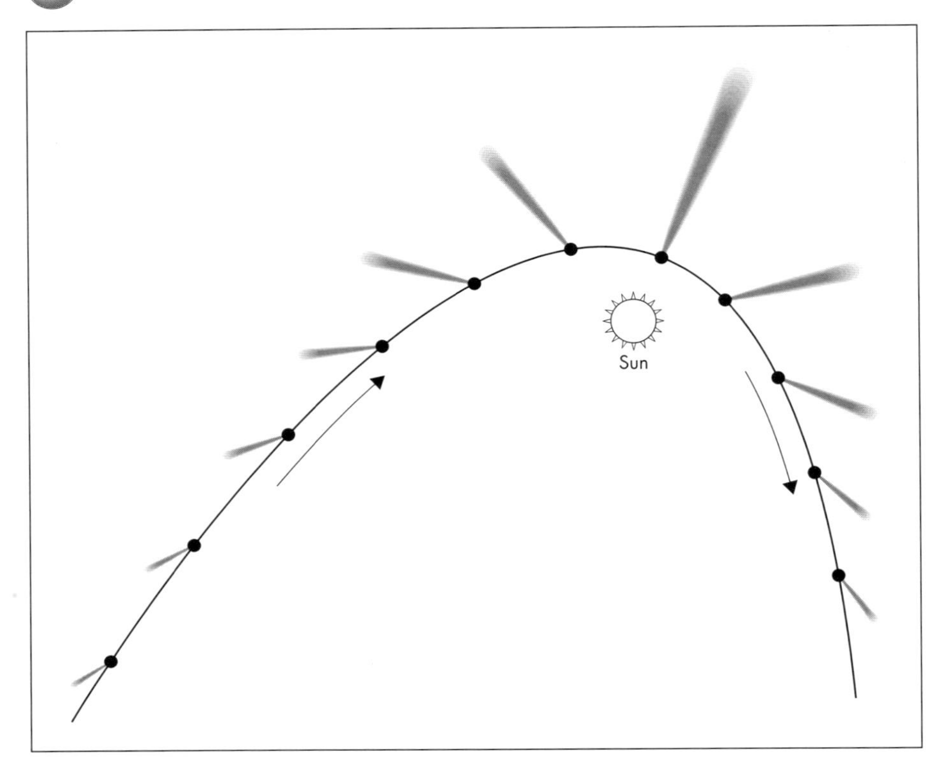

Figure 2.4. Showing how a comet's tail always points away from the Sun.

energy at longer wavelengths. This *fluorescence* is a common phenomenon in nature and has often been exploited by man. For instance, it is the additive in many washing-powders that cause the washed clothes to fluoresce with a slight bluish radiance in sunlight. This gives the impression that the clothes are cleaner and brighter than they really are! In the case of the comets, this fluorescence causes their type I tails to glow with a bluish colour.

Type I tails always point away from the Sun, deviating no more than a few degrees from the anti-solar direction. The consequence of this is that the tail follows the comet when approaching perihelion – but the comet actually moves tail-first after perihelion! (see Figure 2.4).

Type I tails are usually rather straight. However, they are far from being formless. Waves, like the ripples in a flag in the breeze, and bright knots often form in them. These features are seen to slowly work their way along, or through, the tail.

In fact, the analogy with the ripples of a flag in the breeze belies the clue to what causes these features, as well as explaining why the tail always points away from the Sun. The cause lies with the Sun, itself.

The Sun sends out electrified particles streaming radially into space: the *solar wind* referred to earlier in this chapter. These particles carry intricately confused magnetic fields locked-in with them. The ions in a comet, being electrified particles, react to their relative motions with the magnetic fields in a way such as to try to reduce that relative motion. Hence the cometary ions are also forced in a direction radially away from the Sun.

By studying photographs of comets in order to calculate the accelerations of transient features, astronomers have deduced that the forces acting on the ions in a comet due to the solar wind are of the order of a hundred times greater than the size of the Sun's gravitational force acting at the distance of the comet. Once the ions are created in the coma, this "magnetic coupling" mechanism is responsible for selectively carrying them away into the comet's type I tail.

When a comet is far away from the Sun it is moving in a direction which is approximately radial to the direction of the Sun. As it passes perihelion, though, the comet has a significant transverse motion through the solar wind. This can cause a type I tail to develop a slight curvature. However, type II tails are usually very much more curved, and especially so when the comet is passing through perihelion. The same principle applies to the smoke rising from the funnel of a moving steam locomotive. In that case it is the motion of the locomotive through the atmosphere that gives rise to the curved column of smoke.

Type II tails are different in other ways. As well as being generally more curved, they are often much broader. While the type I tails appear bluish in colour, type II tails are noticeably yellowish, or even brownish-orange. I should say here that these colours are only obvious to the naked eye in the very brightest comets. The recent comets C/1995 O1 Hale–Bopp and C/1996 B2 Hyakutake showed the colours well. Binoculars or a telescope help still further and photography can reveal the colours in even faint examples. You will find many examples of colour photographs and images of comets in the CD-ROM that accompanies this book.

Type II tails show spectra very similar to that of the Sun. In fact, they shine by reflected sunlight. Hence we know that type II comet tails are composed of colloidal sized (a thousandth of a millimetre, or so, in diameter) "dust" particles. Smoke is a common example of a colloid. Hence type II comet tails are often known as *dust tails*. Of course, this dust originates from the nucleus, via the coma.

The large curvature of type II tails indicates that the forces repulsing the dust grains from the coma are only just a little larger than the attractive gravitational forces. The main source of this is *radiation pressure*. The pressure exerted by sunlight drives the particles in a direction outward from the Sun.

It is possible that some of the grains would pick up electrostatic charges, given their environment and the manner in which they are swept from the cometary nucleus. This is speculation but if true then a little of the repulsive force on them will come from their interaction with the solar wind.

Comet tails usually extend to lengths of millions of kilometres. Some comets develop extensive type I tails but only weak type II tails. In other comets the reverse is true. Some comets are "gassy", others are "dusty". Some comets have only faint tails, of either type, and some comets develop practically no tails at all. At the other extreme, rare comets have tails longer than 1 AU!

Occasionally a comet sports an apparently forward-pointing spike, or *anti-tail*. A good example is shown in Figure 2.5. Read many books, especially the older ones, about comets and you will find this mentioned as something mysterious. Actually, there is no mystery. The anti-tail is an illusion caused by us seeing part of the highly curved dust tail projected a little forward of the head of the comet because of our viewing angle as we pass through the plane of the orbit of the comet (see Figure 2.6).

The Hydrogen Cloud

As well as an extensive coma and tails stretching millions of kilometres, a comet nucleus carries along with it a truly vast cloud of hydrogen gas. This feature was unsuspected until its discovery in 1970 by the second *Orbiting Astronomical Observatory* (*OAO-2*) satellite. Observations of the comets C/1969 T1 Tago–Sato–Kosaka and C/1969 Y1 Bennett in wavelengths inaccessible from the Earth first revealed these hydrogen clouds. The main source for this hydrogen is explained by the photo-dissociation of the hydroxyl radical in a comet's coma. The very low-massed

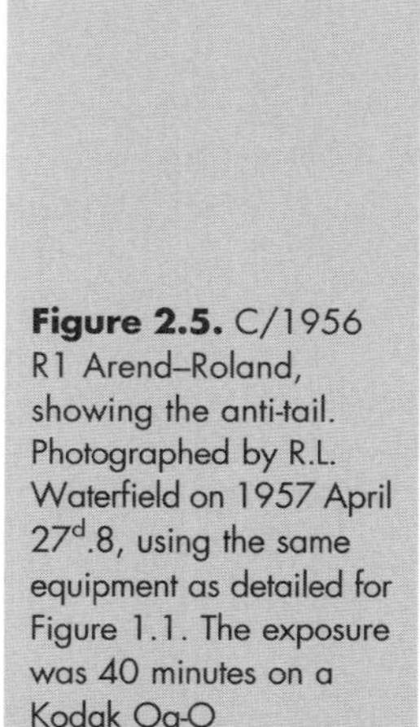

Figure 2.5. C/1956 R1 Arend–Roland, showing the anti-tail. Photographed by R.L. Waterfield on 1957 April $27^{d}.8$, using the same equipment as detailed for Figure 1.1. The exposure was 40 minutes on a Kodak Oa-O photographic plate.

hydrogen atoms created have a very high kinetic velocity (the Kinetic Theory of Gases in action, once again) and the gas streams away from the coma as a vast expanding cloud. This cloud is detectable out to 30 million kilometres, and more, from the nucleus.

Cometary Debris

A comet can survive many passages through perihelion. Thirty apparitions of Halley's Comet have been observed in recorded history, for instance. One of the main reasons for a comet's longevity is the fact that the very sublimation process that removes material from the surface of the nucleus also tends to keep the rest of it much cooler than you might expect.

Try dipping a finger into methylated spirit, or some other volatile liquid, then lifting it out and blowing it. Your finger feels cold. The reason is that the forced evaporation of the liquid removes heat – the latent heat

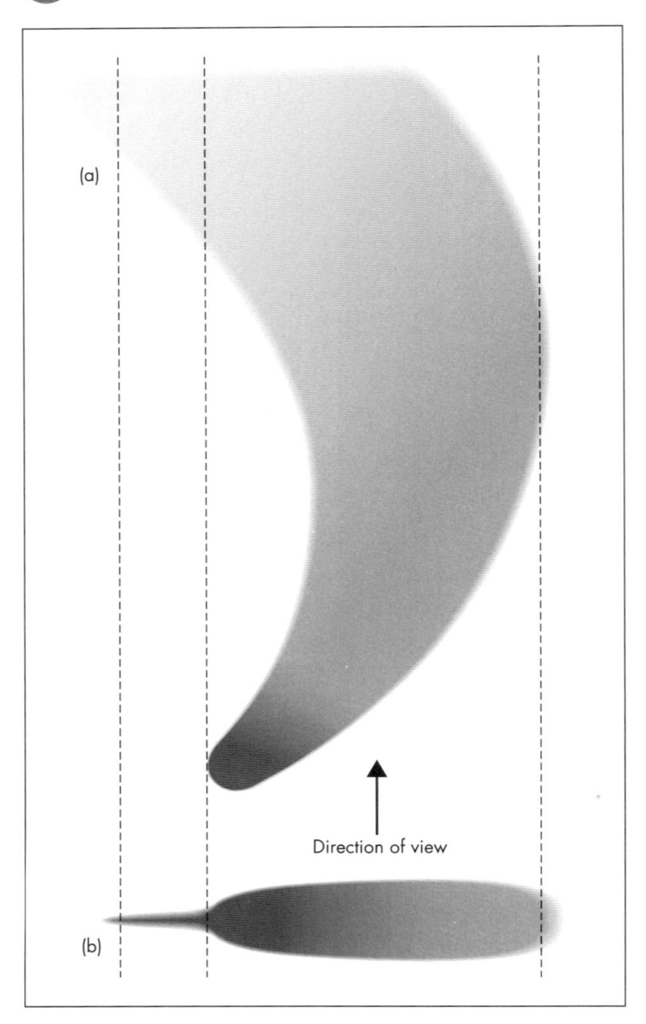

Figure 2.6. Showing how the illusion of an anti-tail arises.

of vaporisation of the liquid. The same effect helps to keep comet nuclei refrigerated. In this case the latent heat removed is very large as the ices sublime to vapour (missing out the intermediate liquid phase because of the low pressure environment).

Nonetheless, the appearance of a coma and a tail indicate that a comet **is** loosing some of its storehouse of matter to the interplanetary medium. The comet will gradually be whittled away with every perihelion passage.

Some comets do not survive their passage through perihelion. One of the most celebrated instances concerned the comet recovered by Captain Wilhelm von Biela in 1826. This object was first discovered in 1772 by Charles Messier and had a period of 6.7 years. However, it was a rather unremarkable object and many of its subsequent returns were missed. Friedrich Bessel predicted that this object should appear again in 1826. Bessel alerted many of his contempories, including Biela.

Biela was first to recover the object, on 27 February 1826. From careful observations he also confirmed its period and it became known as Biela's Comet (under the modern scheme 3D/Biela). It was one of the *Jovian family* of comets, meaning that the comet had been previously "captured" by the gravitational field of the planet Jupiter and its original path had been altered. Subsequently the comet moved around the Sun with a much smaller orbit. You will find more about "captured" comets and comet families in Chapter 6.

The comet was recovered again on its next apparition, that of 1832. Conditions were unfavourable for its next return, in 1839, and so it was not seen. It was predicted to next reach perihelion in 1846 and it was recovered again near the end of 1845. However, astronomers received a shock – they saw two comets flying side-by-side through space! The amazed astronomers watched the comet throughout its apparition and eagerly awaited its next return. In 1852 it, or rather they, were recovered once more. The distance between the comets had increased. Having realised that the original comet had split into two pieces, astronomers eagerly awaited the next returns. Nothing was seen in 1859 but the conditions were very unfavourable. The situation was very much better in 1866 – but still nothing was seen. Still no Biela's Comet(s) in 1872. However, something did occur on November 27 that year: a spectacular meteor shower.

November 27 was the date that astronomers had calculated that the Earth would pass through the orbit of Biela's Comet. This confirmed the widely held suspicion of a link between meteor showers and comets. It was reasoned that debris shed by a comet would spread around its orbit. If the Earth intersected the comet orbit then the particles would produce shooting stars when they entered the Earth's atmosphere. Over the years, the shower has lost its vigour but a few Beilid meteors may still be seen in late November every year.

Astronomers reckon that Biela's Comet strayed too close to the planet Jupiter in 1842 and was disrupted by the strong tidal forces. Clearly, the nucleus of the comet was, at best, very weakly bound. This may be the case with many comets and since this instance, many other comets have been seen to split into two or more pieces.

The Earth encounters many meteor showers each year and most of these have comets as their progenitors. For instance, The Leonids, in mid-November each year, are associated with comet 55P/Tempel–Tuttle and the Aquarid shower of May is associated with none-other than Halley's Comet.

Impact!

Eugene and Caroline Shoemaker, together with David Levy, were undertaking a systematic search for comets using "old fashioned" photography with the 0.46 m Schmidt camera at Mount Palomar. Their ninth jointly discovered comet was particularly intriguing.

They were sure that it was a comet but their discovery image showed its brightest part to be strangely elongated, rather than point-like. Other telescopes were turned towards the object and revealed it to be a comet sporting a nucleus which was fragmented into a number of pieces. When its orbit was determined accurately enough, astrometrists could back-track their calculations to investigate its past history. This showed that the comet had made a very close fly-by of the planet Jupiter in July 1992.

The comet was disrupted by the tidal forces imposed on it as it flew past the massive planet. When astrometrists forward-tracked their calculations they received a big shock: this time around the comet would not merely fly past Jupiter – it was going to collide with the planet!

Professional and amateur astronomers watched with growing excitement as the separate nuclei of comet Shoemaker–Levy 9 (properly D/1993 F2 Shoemaker–Levy 9) evolved and inexorably moved towards the great planet Jupiter. Some of the fragments were very small. Some even vanished altogether, showing that they must have been very insubstantial in nature. Some further subdivided. Altogether there were about 21 members, the number varying from time to time, variously strung out like pearls on a cosmic necklace as they headed unknowingly towards their fate (see Figure 2.7).

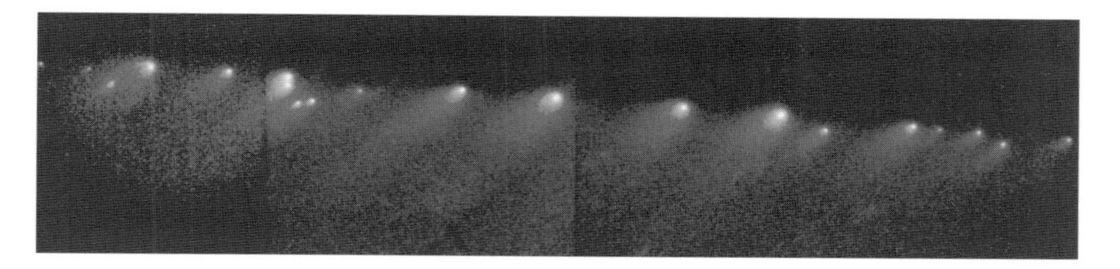

Figure 2.7. The pieces of the disintegrated comet D/1993 F2 Shoemaker–Levy 9 on their way towards impact on the giant planet Jupiter. The Hubble Space Telescope took this image in January 1994.
Courtesy NASA and STScI.

Under the spotlight of the international media, astronomers witnessed the first of the fragments slam into the great planet at 60 km/s on 16 July 1994. The carpet-bombing continued for the next six days.

Although, to the frustration of everybody, most of the impacts occurred just out of sight, a little round the limb on the rear of the planet, infrared telescopes such as UKIRT on Hawaii usually saw the flash of each impact followed by a great "fireball" and a plume of gases rising from the impact site. Each explosion had an energy vastly exceeding the sum total of the world's entire nuclear arsenal.

Not knowing what we would see next, we all carefully scrutinised Jupiter. The planet was, at that time, very low in a twilight sky as seen from the UK and it was out of view of my largest telescope. I had to make do with the views of the planet using my old 158mm Newtonian reflector. Given the apparently flimsy and insubstantial nature of the Comet Shoemaker–Levy 9 fragments of nuclei, some commentators asserted that nothing significant would be seen after the impacts – and that as the impacts themselves were out-of-view, the whole thing would be something of a non-event.

I didn't agree with this opinion but I was as shocked as anybody else to see huge black "scars" from each impact site rotate into view. Despite the poor seeing that attended the low altitude of Jupiter and small size of the telescope, these "scars" were very easy to see (see Figure 2.8). Most of them lasted more than a month, changing their size and shape all the time. The impacts all occurred within a little of 44°S latitude on the planet. By the time Jupiter was recovered again after its conjunction with the Sun the individual "scars" had become stretched around the planet and had merged into a new temporary belt, lasting about a year. Though comet Shoemaker–Levy 9 is now gone, it will never be forgotten. We learned much about the complexities of Jupiter's cloudy mantle and about comets from this once-in-a-millennium event.

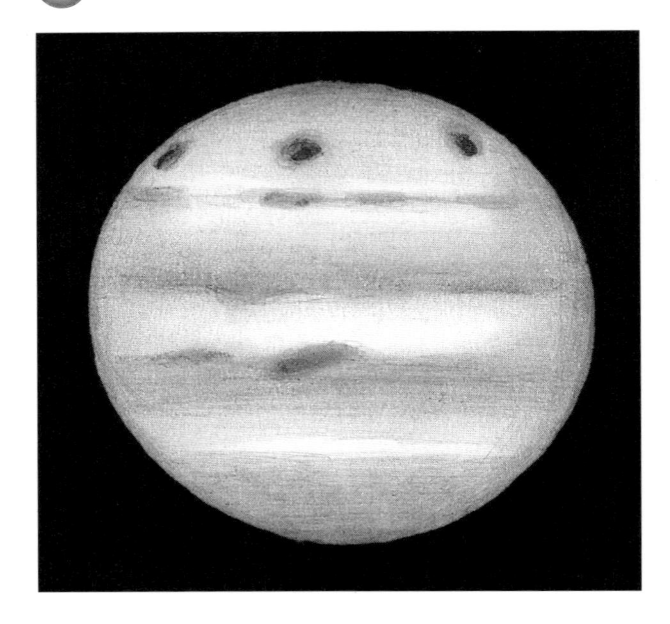

Figure 2.8. The impact scars of comet Shoemaker–Levy 9 on Jupiter, as observed by Gerald North on 1994 July 20^{d} 20^{h} 21^{m} UT. The marks were very obvious even though the seeing was very poor and the observer was forced to use his 158 mm Newtonian reflector because the planet was too low down for access with his larger telescopes.

If Jupiter can be hit by a comet, what about the Earth? Well, calculations show that we owe a dept of gratitude to Jupiter. It actually protects us from impact by virtue of its great gravitational field. It goes around the Solar System like a celestial vacuum cleaner, reducing the chances of the Earth getting hit by anything substantial. However, it cannot protect us completely...

On the morning of 30 June 1908 a brilliant fireball was seen to hurtle across the sky over Siberia. It terminated in a thunderous explosion that was heard over a thousand kilometres away. Seismometers all over the world recorded shock-waves and peculiar atmospheric effects occurred in the days that followed the explosion.

Scientists fixed the explosion site as being in the remote forests of the Tunguska region of Siberia. Later investigators were stunned by what they found: over 2000 square kilometres of the forest had been decimated, with about 1000 square kilometres of the central region largely incinerated. The trees were knocked flat in a radial pattern and stripped of all foliage for several kilometres around the central region.

The investigators expected to find a crater, the usual signature of a meteoritic impact, or at the very least some fragments of meteorite. None was found. Predictably, the cranks and crackpots had a field day, variously proposing all sorts of wild theories – alien spacecraft, antimatter missiles, etc. – to explain the fireball and explosion. However, the more sober of the scientists proposed the icy nucleus of a comet as the probable cause.

Recent expeditions have had the advantage of utilising modern investigative techniques. Traces of certain isotopes have been found in peat samples and these tend to confirm the comet hypothesis. However, other samples reveal traces of elements that one would expect from a rocky body. Recent calculations also show that a solid body, such as a piece of asteroid, can explode while still in the air and can leave no crater or sizeable chunks to pepper the ground.

So, was it a solid meteorite (perhaps a chunk of asteroid) or a comet nucleus that caused the Tunguska explosion? We have to admit that we are still not sure, though the evidence does slightly favour a meteor. Various mathematical models suggest that the object exploded while it was still about 6 km above the ground. The magnitude of the explosion is reckoned to be about 2000 times that of the Hiroshima atomic bomb. Knowing the energy, we can estimate the mass of the projectile. It turns out to be about 100 000 tons.

Whatever the cause for the explosion, it was lucky that Tunguska was relatively unpopulated and so casualties were few. Even so, the effects of a comet nucleus of more typical mass, at least tens of thousands of times greater, hardly bears thinking about. The largest comets have nuclei several tens of **millions** of times larger than that of the Tunguska projectile. Surely nowhere on Earth could have escaped the devastation that would have followed such an impact?

Naturally, our thoughts turn towards the global mass extinctions of the past aeons. Were comet strikes responsible for any of these? Maybe. Maybe, though, comet impacts have actually favoured life on this planet. Some scientists have speculated that life itself was first initiated in space and it was brought to Earth via comets. Some also assert that pandemic diseases also arrive from time to time by this method. Not many others support this view. However, where did the water in our oceans come from? Some of it surely originated from the quota of water that occupied our place in the primordial Solar System. It is also quite likely that much of it was delivered onto the Earth by impacting comets. In fact, it would require only about one comet strike every thousand years for the first billion years of the Earth's history to fill our oceans. In those times when the Solar System was chaotically full of debris one comet strike per millennium would be a very conservative estimate. The idea of life being seeded on our planet by comets might be unnecessarily fanciful, but life-giving water being supplied by comet impacts certainly is not.

Rendezvous

Before the recent great comets Hyakutake and Hale–Bopp, if you asked anyone in the general public to name a comet they would probably say "Halley's Comet". Now that the memories of Hyakutake and Hale–Bopp are fading, if you tried the same question on members of the public again you would probably get 'Halley's Comet' as the answer once more. Undoubtedly Halley's Comet is the one with the enduring fame.

Much "Halleyballo" heralded its last apparition in 1986. This comet received unprecedented attention from amateur and professional astronomers alike. However, it cannot be said that the apparition was at all favourable from the Earth's northern hemisphere, though people living in the Earth's southern hemisphere did get a better view (see Figure 2.9).

Figure 2.9. Halley's Comet photographed from the southern hemisphere in 1986. Image courtesy Brian Carter, South African Astronomical Observatory.

Halley's Comet was generally very much better seen in 1910. In spite of the absence of modern electronic detectors and computers at the time, photography was well advanced by then. As a result, the comet has been intensively studied during the entirety of both of its apparitions of the twentieth century. We have learned a lot from 1P/Halley.

Although the 1986 apparition of Halley's Comet was a disappointing one from the point of view of the Earth-based observer, it did mark the first really close-range examination of any comet owing to a fleet of space-probes dispatched by us curious humans.

Years of planning went into the endeavour. In the event six probes were sent, the first of these being *Vega 1* and *Vega 2* launched in December 1984. They whizzed by the nucleus of Halley's comet, at a distance of about 8000 km from it, on 6 March and 9 March respectively. Next to launch a probe to the comet were the Japanese. Their *Sakigake* probe was sent aloft in January 1985 and passed within 4 million km of the nucleus on 11 March 1986.

Next in the chronological catalogue of launches came the European Space Agency (ESA) probe *Giotto,* in July 1985. It passed the comet nucleus at very close range (596 km) on 14 March 1986. Meanwhile, another Japanese probe, *Suisei*, was dispatched in August 1985, flying past the nucleus at a distance of about 200 000 km on 8 March 1986.

Last, but by no means least, was the *International Cometary Explorer* (*ICE*). One of the most versatile of the space probes ever sent aloft, it was originally launched in December 1978 to study the solar wind. NASA decided to make use of this probe to study comets (and adopted the moniker *ICE* for it). NASA controllers used the gravitational influence of the Moon to swing the probe towards a new target: the comet 21P/Giacobini–Zinner. It passed through the tail of this comet in September 1985, then onwards to Halley's Comet, which it passed in 1986. Admittedly it did not fly past Halley's Comet at close range, only closing in to about 40 million km of the nucleus on 28 March 1986 but it still provided much valuable information and offered a comparison between two comets.

Giotto took the prize for producing the most spectacular results. It passed through the inner coma of Halley's Comet and took many colour pictures while it was also busy taking many physical measurements of its environment. The velocity of the probe relative to the dust particles was of the order of 70 km/s. The designers had allowed for this by including a double-layered shield to fend off the damaging effects of the hail of particles. Even so, a larger-than-average particle did hit the spacecraft just seconds before closest approach and jarred it off-line sufficiently that effective radio communications were lost for about an hour. Even so, the approach data was very instructive and some spectacular pictures of the nucleus were obtained (see Figure 2.10), showing the jets of gas and dust in action erupting from the nucleus.

Figure 2.10. The nucleus of Halley's Comet imaged at close range on 1986 March 14. Notice the two jets issuing from the nucleus and the fact that the "night-side" of the nucleus can be seen silhouetted against the dust in the coma extending behind it. This is a composite of 68 images taken by the Halley Multicolour Camera onboard ESA's *Giotto* spacecraft. The copyright for this image lies with the Max-Planck-Institut für Aeronomie, Lindau/Harz, FRG and this image is reproduced here by special permission. The authors are grateful to Dr H.U. Keller for supplying this image.

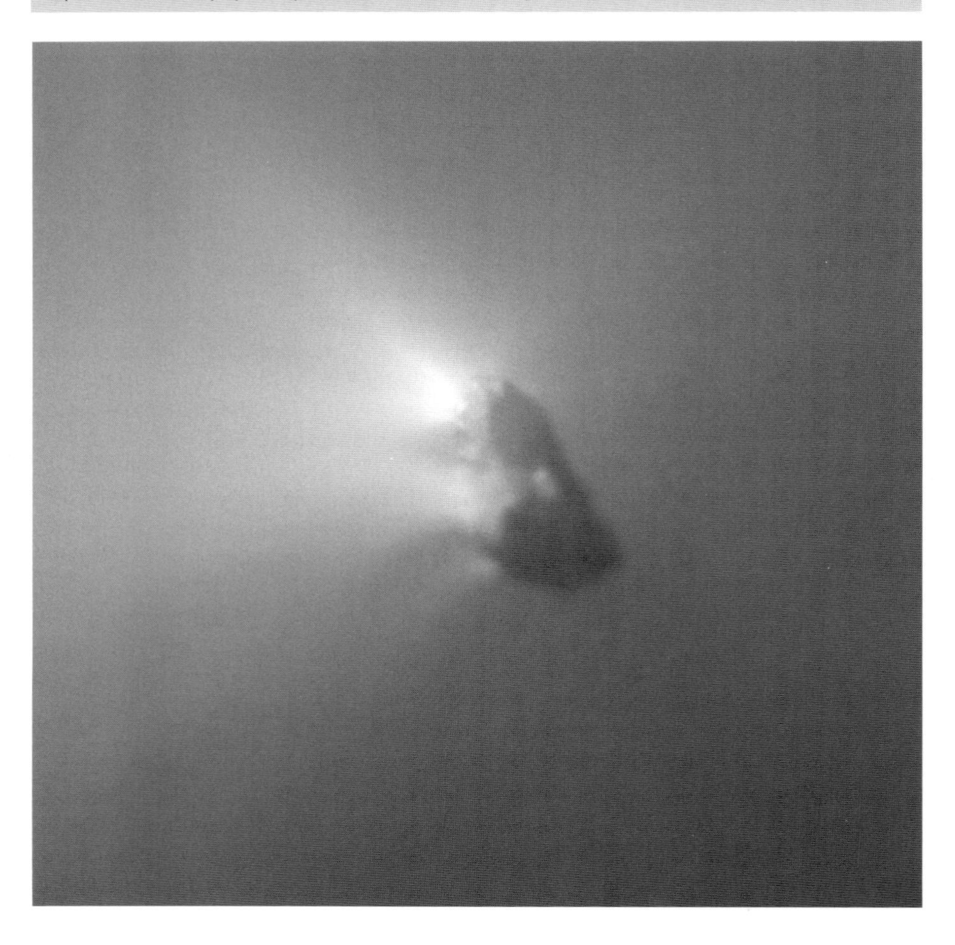

The images showed the nucleus of Halley's Comet to be an extremely dark and potato-shaped body spanning about 8.2 km $\times$ 8.4 km $\times$ 16 km. Gravimetric measurements indicated that the average density of the nucleus ranges about 500 kg/m^3. This is about half the density of normal ice and was a surprise at the time. It must mean that the nucleus is a very loose agglomeration of ice crystals and any rocky core must be very insubstantial. Since then, astronomers have begun to realise that many other cometary nuclei may have even smaller densities – since many have been easily fragmented by tidal forces, as described earlier.

The dark covering was also unexpected, and is still not definitely explained. Actually, its *albedo* – the amount of the incident light reflected – is a mere 2.7 per cent. This is **very** black, indeed! Perhaps it is the result of intrinsically dark materials. If so, what are these materials? Is this darkening caused by the action of radiation on the surface volatiles? Could, then, carbon or carboniferous compounds result from partial photodissociation? Many outer Solar System objects have very low albedos. Is this a clue, or just a red herring? *Giotto* has raised as many new questions as it has solved past mysteries.

To help solve these new questions many more spacecraft are being sent to explore comets. In 1998 an experimental spacecraft called *Deep Space 1* was launched to test various space technologies. At the end of its primary mission NASA approved an extended mission that sent it to within 2300 km of the nucleus of 19P/Borrelly. It passed the comet on 22 September 2001 and sent back some spectacular images of the 8 km-long bowling-pin shaped nucleus (see Figure 2.11). These showed a variety of surface features ranging from smooth bright plains to dark mountains and faults.

Whence Come The Comets?

We can only see a comet when it approaches the inner realms of the Solar System. Nonetheless, they can be tracked and their orbits calculated to include the part, actually the major part, that we cannot see. It is in this way that we know that the majority of comets come from well beyond the outermost known planets of our Solar System.

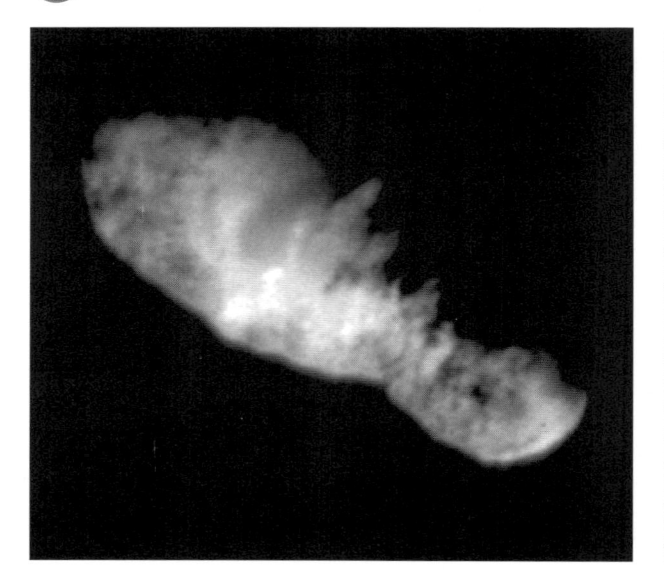

Figure 2.11. The nucleus of comet 19P/Borrelly imaged from the *Deep Space 1* probe on 2001 September 22. Image courtesy NASA/JPL/Caltech.

By the middle years of the twentieth century it was realised that the lifetimes of the observed comets were very short compared to the lifetime of the Solar System, and yet there seem to be a fairly steady supply of pristine comets arriving for their first perihelion passage each year. Clearly there is some vast reservoir of comets.

Thanks to the late *Jan Oort*, who was an expert on the modelling of the dynamics of complex gravitational systems, an idea has evolved that explains the origins of the comets. We now think that comets reside in a huge, shell-like, cloud, now known as the *Oort Cloud*, which extends to over 50 000 AU from the centre of our Solar System. The Oort cloud contains something like 200 thousand million comets in deep freeze. Their total mass comprises something around 6×10^{23} kg, or about one tenth the mass of the Earth.

Out there these frozen ice-balls experience tidal forces caused by the motions of the other nearby stars in our galaxy (in our region of the galactic disk the stars have average separations of about 4 light years). These forces are a significant fraction of the gravitational pull that keeps the ice-balls gravitationally bound to the Sun and so have a marked perturbing effect on them. Some even topple from their precarious orbits and plunge in towards the Sun. In fact, the continuing process produces a veritable "rain" of comets that showers the inner Solar System.

It seems very likely that comets were formed from the condensing protosolar cloud that eventually gave rise to the Sun and planets of our own Solar System. In that way comets provide a direct link to the processes that gave rise to the Sun and planets, and even us! If our ideas are correct, then comets provide examples of the earliest, virtually unprocessed, materials from which our Solar System originated.

There is a class of icy, comet nucleus-like, objects that reside in rather eccentric and generally highly inclined orbits at around the distance of the eighth planet from the Sun. Some of these are now known to be several hundred kilometres in diameter. These were formerly known as *Kuiper Belt* Objects (and more properly that should be *Edgeworth-Kuiper Belt Objects*). Now, the fashion dictates, they are known as *Trans-Neptunian Objects* (*TNO*s). Indeed, the ninth planet, Pluto, is now thought by many to be the largest of these TNOs!

There are other comets we see that stay within the planetary realm of the Solar System. How do all these fit into the scheme of things? Actually, the whole subject of comet orbits is very involved. For instance, the major planets of the Solar System have an important part to play in the lives of many comets, as well as the death of a few of them. You will find much more about comet orbits later in this book. You will also find more information about the physical and chemical natures of comets in some of the sources listed in the Appendix. For now, though, space within this book is pressing. So, we must conclude our quick-look survey of the physical and chemical natures of comets and turn our attention to the main theme of this book – the actual process of observing these celestial phantoms.

Chapter 3 Comets in Vision

There is something about actually seeing a comet, or for that matter any celestial body, with your own eyes (whether or not it be via reflective and refractive optics) that surpasses merely looking at any photograph or CCD image of it.

At very irregular intervals, generally every few years, a comet arrives in our skies that is bright enough to be seen by any normal-sighted person without any optical aid. Arming ourselves with binoculars allows us to potentially increase the number of comets we can observe to several per year because we can then observe fainter ones – and there are **many** more faint ones than there are bright ones. A telescope of greater light grasp than the binoculars will help us to see yet more comets, of course. Every extra bit of optical aid potentially brings a great increase in the number of comets observable, as well as helping us to see more physical details on any comet however bright or dim it might be.

Telescopes for Observing Comets

A reasonable question you could ask is: "What is the ideal telescope I could choose for observing comets?" Unfortunately, there is no straightforward answer. A comet that appears as an excessively faint misty patch spanning a mere few arcseconds would be best appreciated with a fairly large magnification applied to a large aperture telescope. Perhaps $\times 200$ to $\times 300$ on a 0.5m (or even larger) reflector would be a good choice, if you can afford a telescope that big. A similar magnification and large size of telescope might be best for observing certain details, such as jets and hoods, in the coma of even a bright and extensive comet. Yet, to frame the same comet complete with much of its tail would need a field of view of many degrees wide; a challenge even for binoculars.

Moreover, even a comet that is large and bright at its best will have started (once it becomes visible at all) as a small, faint, fuzzy patch before it reaches perihelion and will diminish to that again as it recedes afterwards. So, one might desire to have a whole arsenal of types and sizes of equipment to suit all possibilities. That is quite impossible for most of us. The "silver lining in the cloud" is that you can gain some satisfaction and a whole lot of pleasure, and even perform some scientifically useful tasks, whatever the equipment you possess.

I recommend adopting a positive attitude of mind and press whatever equipment you have into service. Indeed, you can always, by making a few judicious additions, optimise your existing equipment to get the best performance from it.

At this point I should mention that there are a number of books providing information about astronomical equipment. *Astronomical Equipment for Amateurs*, written by Martin Mobberley and published by Springer-Verlag in 1999 is a good general book covering the range of equipment on the market at the time the author wrote it. *Advanced Amateur Astronomy*, written by one of the authors of the present book (North) and published in its second edition in 1997 by Cambridge University Press, includes detailed advice and information concerning the selection, setting up, evaluation and testing, of astronomical equipment (including necessary adjustments) and their adaptation for a range of observational tasks, including comet observation.

Here, since space is at a premium, I must assume that you have a good general knowledge about equipment for amateur astronomers, or refer you to other works on the subject, and just deal with a few matters that are specific to the needs of the comet observer.

One of these matters is the interrelated, and somewhat convoluted, subject of telescope aperture and magnification and how these factors relate to the perceived brightness of the viewed image. This is extremely important whatever equipment you press into service for your comet observing and is perhaps the most important factor to bear in mind when selecting new equipment. So that is where we will begin.

Image Brightness, Magnification and Telescope Aperture

A star produces a *point image*. The perceived brightness of this point image is determined almost entirely by the aperture of the telescope. Focal ratio has no effect at all.

Most practical observers agree that if you observe from a good, dark, site and experience a night sky of excellent transparency then you will see stars of the 6th magnitude. Indeed, if your visual acuity is even slightly better than the average then you ought to be able to see stars as faint as $6^{m}.5$ on such a night from such a site. Use a telescope with an appropriate magnification and you will see stars which are much fainter than that. How faint? Well, that is a matter of some conjecture. There are a number of old predictive formulae. These give widely different results.

However, Bradley E. Schaefer of the NASA–Goddard Space Flight Center has conducted a wide practical survey of contemporary telescope users. He finds that provided the magnification is high enough (perhaps $\times 200$ for a 150mm aperture telescope, increasing to perhaps $\times 300$ for a 400mm aperture) one can do rather better than most of the older formulae predict. One of the authors (North) has proposed a formula which fits the results of Schaefer's survey very well:

$$\text{Practical Limit } m_v = 4.5 + 4.4 \log D,$$

where D is the telescope aperture, in millimetres, and m_v is the stellar limiting magnitude attainable. Any formula such as this can only ever be a rough guide because there are many factors which will affect the final result. However, based as it is on a real practical survey of modern-day telescope users, it should be a better guide than most of the older ones. Table 3.1 shows the stellar limiting magnitudes achievable with a selection of telescope apertures, as predicted by this formula.

The reason for the magnification being important is that the star image is seen against the image that the

Table 3.1. Telescope limiting stellar magnitudes

Telescope aperture (inches/mm)	Limiting magnitude
6 / 152	14.1
8 / 203	14.7
10 / 254	15.1
12 / 305	15.4
14 / 356	15.7
16 / 406	16.0
18 / 457	16.2
20 / 508	16.4

telescope produces of the sky background – **and this background is not perfectly black.**

The sky background behaves like an *extended image*, meaning that the light from it is spread out, as opposed to being concentrated into a tiny point. The telescope gathers an amount of light determined by its aperture and presents this in the image that we see when we look through the eyepiece. Use a given magnification and the star will be seen as a point of light. Even from a site not bedevilled by light pollution, the surrounding sky will appear as a grey wash of light covering the field of view. Double that magnification, though, and the light in a given patch of sky background will be stretched to four times its previous area and so will appear only one quarter as bright as before. In general the brightness of the sky background is inversely proportional to the square of the magnification.

Using more magnification effectively darkens the sky background while the star image is little affected (until too much magnification is applied, when its image ceases to be point-like). The better **contrast** between the star and its background at the higher magnification makes it **seem** brighter. Image contrast is a crucially important factor when trying to see things. Hence, up to a point, increasing the magnification used on a given telescope makes the stars we see through it appear more distinct. It also enables one to detect stars that may be lost in the background sky glow at lower magnifications. Thought of in this way, Schaefer's results hardly seem surprising.

What magnification should you choose when observing a comet through a telescope? Given the foregoing, and since a comet forms an extended image, surely the lowest possible magnification is best in order that the light from the comet is most concentrated? Well, I am

Figure 3.1. The bright image of the primary mirror of Gerald North's 0.46 m Newtonian reflector can be seen against the eye lens of the eyepiece. This is the exit pupil and is the best position for the observer's eye pupil in order to see all of the light gathered by the telescope. The exit pupil is centred on the optical axis but is formed a little way in front of the eye lens, which is why it appears displaced to the upper-left in this oblique view.

afraid that things are not quite that simple. There are other factors to be taken into account.

The first of these is that there is a limit to the gain in image brightness we see as we lower the magnification we use on a given aperture of telescope. The eyepiece of a telescope produces a disk of light, usually known as the *exit pupil*, or the *Ramsden disk*, or sometimes as the *eye ring*, which is, in effect, the image of the telescope objective formed by the eyepiece. The exit pupil is situated just a little distance, known as the *eye relief*, from the eye lens of the eyepiece (see Figure 3.1). All of the light from distant objects which is collected by the objective of the telescope passes through the exit pupil. The observer automatically steers his eye pupil to coincide with the exit pupil in order to receive the maximum amount of the light collected by the telescope.

The size of the exit pupil is inversely proportional to the magnification the eyepiece produces with the telescope. In fact, we can calculate the size of the exit pupil from the following relation:

$$\textit{diameter of exit pupil} = \textit{aperture}/\textit{magnification}$$

The diameter of the exit pupil and the aperture of the telescope have to be given in the same units. As an example, if a 200 mm reflecting telescope has an eyepiece plugged into it to produce a magnification of ×40, then the exit pupil produced has a diameter of 5 mm.

The eye pupil of a teenager can expand to about 7 mm or 8 mm when fully dark adapted. As one ages this figure decreases. The eye pupil of an average 30-year-old person will expand to around 6.5 mm and this figure decreases further by about 0.5 mm per decade. Thus, an average 60-year-old has an eye pupil that will open to only about 5 mm at best.

If the pupil of the observer's eye is smaller than the size of the exit pupil then not all of the light collected by the telescope can enter his eye in one go. Putting it the other way round: the observer must choose a magnification in order to make sure that the exit pupil is no bigger than the his eye pupil if he is not to waste some of the light grasp of the telescope.

Consequently, there is a minimum magnification you should use on a telescope. Use a lower magnification than that and you will waste some of the precious celestial light collected by it. If you are a 60-year-old and are using a 200 mm aperture telescope then you will be wasting some of its light-grasp if you use an eyepiece with it giving a magnification of less than ×40.

What does "wasting light" mean in practice? Obviously, anything in the image that is point-like, for instance the image of a star, will seem less bright than it should. Indeed, the faintest point images might be then rendered too dim to see at all. It is just as if the telescope has had its aperture reduced, or it has been *stopped down*, as we say.

The effect on an extended image is a little different. Let us examine what happens by means of an example. Let us say we are sixty years of age (and so have an eye pupil of 5 mm diameter) and are using our 200 mm reflector with a magnification of ×40. Our eye pupil just matches the size of the exit pupil produced by the telescope–eyepiece combination. If we carefully steer our eye into position (and hold it still enough) then all of the celestial light gathered by the telescope will find its way into our eye. We see our comet as a large fuzzy patch in the centre of the field of view.

Since it is not one of the more spectacular comets, we decide it is a good idea to concentrate the light further.

We reach for another eyepiece. That one gives a magnification of ×20 with the telescope. We excitedly look through the eyepiece – and are disappointed at the result!

Instead of seeing the comet half as big and its dim light concentrated to make it appear brighter, it appears not one jot brighter than it did with the higher magnification. Indeed, thinking that the same amount of light is now concentrated into half the linear size, and therefore one quarter of the area, could mislead us into thinking that the comet image should appear four times as bright. Yet, it does not.

What has gone wrong? The answer is that choosing half the magnification now makes the exit pupil twice as big as before – and so (in our example) twice the diameter of our eye pupil. Put another way, the exit pupil now has four times the area of our eye pupil. The light gathered by the telescope is spread across the exit pupil (and remember, this is effectively the image of the telescope objective formed by the eyepiece – it is **not** the focused image of the celestial body itself). By intercepting only one quarter of the area of the exit pupil, our eye is admitting only one quarter of the light gathered by the telescope.

We said that the image of the comet seen at ×20 has only one quarter of the area it does at ×40. Since the image of the comet now formed on the retina of our eyes only utilises one quarter of the light gathered by the telescope at ×40, the end result is that the amount of light energy per unit area of the comet image formed on our retina is just the same as before! Lowering the magnification has not increased the *apparent surface brightness* of the image of the comet at all. The only change has been to make the comet appear smaller.

Looking at this phenomenon from another angle, we could fit a disk with a central hole in over the objective of the same telescope, so stopping it down. With the applied magnification of ×40 we would be proportionately reducing the size of the exit pupil from its initial 5 mm diameter. The image our eye can see (with its 5 mm eye pupil) would get progressively dimmer if we substituted stops of smaller, and smaller, clear aperture.

Do the same with the eyepiece plugged in that gives ×20 and we will see no change until we have reduced the diameter of the clear aperture of the telescope to 100 mm. At that point the exit pupil produced by it will have a diameter of 5 mm, the same size as our eye pupil.

Subsequent reductions in the clear aperture of the telescope will then have a noticeable dimming effect on the image we see through the eyepiece.

Whatever eyepiece we choose, we cannot do better than a 1 to 5 ratio of magnification to aperture in concentrating the light from the comet and presenting it to our eye with its 5 mm diameter eye pupil. Using a larger telescope cannot help us improve this ratio – it is limited by the physical size of the pupil of our eye. If our eye was a youthful one with a 7 mm pupil diameter then, and only then, might we see any benefit from using a lower magnification. The ratio of minimum magnification to aperture would then be 1 to 7.

Physiology also has an important bearing. Firstly, the size of an image is very important to how we perceive it. A large dim object is usually much easier to see than a small dim object. We have mentioned the importance of image contrast already. There is also a big difference in the way our brain represents small differences in the brightnesses of very dim objects seen against a dim background, to that when the ambient lighting is much brighter (as is the case for images at lower magnifications).

Since it is the **actual practice** that really counts, I will not waste precious space here by going further into the physiology of vision. Instead, I will suggest that you perform your own practical trial. I will give the results I expect you to obtain – and this result will surprise you – but you will really only fully appreciate the consequences for your observations of comets if you try the experiment for yourself. You will then be able to make much more informed choices about the equipment you select/make, and in particular the eyepieces you purchase to use with your telescope, in order to get the best from your comet observing.

The trial is a simple one but does require you to have some form of telescope (it can be a small one) with a range of magnifications usable (either a selection of eyepieces, or a "zoom" arrangement – as is the case on many of the very cheapest small telescopes). You do not even need to wait for a bright comet to make its appearance. There are plenty of objects in the sky – nebulae and galaxies – that will do as "pseudo comets".

Take a bright and fairly diffuse object such as M42, the Great Nebula in Orion. If the night is even halfway decent, you should easily see it with the naked eye. Try and gauge its visual extent and try to sketch, or at least make a mental note of, what you can see in the way of detail. Then look at it through the telescope with the smallest available magnification. Try and gauge its visual extent now (you might have to pan the telescope if the eyepiece field of view is very small) and again note what detail you can see and how easily that detail can be seen. Increase the magnification and repeat the procedure.

Knowing the values of magnification and the size of the telescope objective, you can work out the size of the exit pupil produced for each magnification. Actually, small telescopes are seldom provided with very low magnification (producing an exit pupil of larger than 5 mm diameter) eyepieces, so a pair of low-power binoculars might have to suffice for that part of the exercise.

If you will allow me to predict the results you will obtain, we can take our discussion forward here. I am sure that you will find that if you could arrange a magnification to equal that of your eye's pupil diameter (yes, I know that you can only guess at this value without a further measurement – actually not hard to do if you have a *pupil gauge*) then your view of M42 would be appear bigger simply because of the magnification factor. It may also appear more distinctive (again thanks of the increase in its size). However, I predict that it would still cover the same extent of actual sky (perhaps about a quarter of a degree, or a little more) as when it is seen with the naked eye, the dimmer outer parts then blending into the greyness of the sky background.

With progressively higher magnifications, though, the sky background will be seen to darken and the magnified view of the nebula will appear to cover more width of actual sky (perhaps even a degree, maybe more) before its outer tendrils merge into the background. In other words, you will now be seeing parts of the nebula as distinct that were merged into the sky background before. This is the result that I imagine you will find surprising. You will see that image of the nebula is also generally better contrasted and the details within it are easier to see at this higher magnification.

Go too far with the magnification, though, and the image becomes too diluted and dim. However, you may be surprised just how high a magnification you can use before you start to lose significant amounts of image detail.

If your sky is reasonably dark then I predict that your best view of the nebula (or perhaps just a part of it, since your field of view may be restricted) will be at a magnification that produces an exit pupil diameter of somewhere around 2.5 mm – "best" being defined as seeing the filamentary detail with the greatest ease and having the nebula appearing with its largest angular extent on the sky. This is a magnification equal to about ×0.4 of the telescope aperture in millimetres (about ×10 per inch). If your sky is not so dark then an even higher magnification may prove beneficial to "separate" the nebula from the background glow.

Try the effect of different magnifications on other "deep-sky" objects of differing size and brightness. A faint and diffuse object such as M1, the Crab Nebula, might look best at a lower magnification than ×0.4 per millimetre but a fairly compact and well-defined object such as M57, the Ring Nebula, will look at its best with a little more magnification applied. Try the same test using the great galaxy in Andromeda, M31, and you will find, though, that this very large and amorphous disk looks most impressive with a very low magnification. However, the elusive details in the disk of the galaxy are still most easily detected at higher powers.

Think about the variation of the shapes, sizes, and brightnesses of comets (and the component parts of comets) and you can see why I have devoted so much of the limited space in this chapter to the subject of magnification(s) used on particular telescopes.

The advice usually peddled concerning observing diffuse objects through telescopes is to use the lowest possible magnification. Many authorities like to talk of "equalisation magnification", or "visual equalisation", or some other similar phrase which means using a power as low as possible and having the exit pupil the same size as the observer's eye pupil. This advice is most definitely wrong! **Limit yourself to such a low magnification and you will be throwing away much of the potential performance of your telescope** – but don't just take my word for it; do the trial and find out for yourself.

While never wishing to waste a telescope's light grasp, it is often useful to have the widest field of view possible. For that purpose a "visual equalisation" magnification value **is** useful, even if our chosen comet may not be seen at its best with that particular eyepiece. The widest fields of view come from telescopes of a particular design: the so-called *rich-field telescopes*.

Rich-Field Telescopes

Despite stating earlier that there is no one ideal telescope, if I was pressed very hard to select just one telescope that would be a good choice for the comet observer then I would probably have to say something with an aperture in the range of 100–200 mm and a focal ratio in the range of f/4–f/6. At least that would be a fairly good compromise between aperture and field of view (both of which should be as large as possible, though making one big inevitably makes the other small). This compromise is, though, strongly biased towards the needs of the observer who wishes to use manual means for tracking down comets and those who wish to see the biggest comets (spanning more than a degree – more than twice the diameter of the full Moon – across the sky) to their best advantage.

This type of instrument (at least in sizes of no more than about 160 mm aperture) is often known as a *rich-field telescope*. Many manufacturers market telescopes that come into this category – and some examples are shown in Figures 3.2 to 3.4.

Figure 3.2. The Orion VX114-ED Refractor. Courtesy Orion Telescopes and Binoculars.

Figure 3.3. The Orion SpaceProbe 130ST EQ Reflector. Courtesy Orion Telescopes and Binoculars.

Figure 3.4. The Orion Sky View Deluxe 8 EQ Newtonian. Courtesy Orion Telescopes and Binoculars.

There is a common misconception I ought to dispose of straight away concerning low focal ratios in **visual** telescopes. A telescope with a low focal ratio **does not** produce a brighter image in the eye of the person looking through the eyepiece than would be the case for a telescope with higher focal ratio (and the same applied magnification).

If you were to apply a magnification of ×50 to a 200 mm f/4.5 telescope (perhaps a Dobsonian reflector) and used it to observe a comet you would see the same apparent image brightness as you would if you compared it with the view through a 200 mm f/10 telescope (perhaps one of the popular Schmidt–Cassegrains) with a particular eyepiece plugged into that one chosen also to give a magnification of ×50.

You might then ask: "If the comet doesn't look any brighter in the low f-number telescope, why should there be any preference for using one?" The real reason for using an instrument of low focal ratio is that it has the potential to provide a much larger apparent field of view.

The image-scale, expressed in arcseconds per millimetre, at the principal focus (no additional optics) of a telescope is given by:

$$\text{image scale} = 206265/f,$$

where f is the focal length of the telescope, in millimetres. As an example a 150 mm f/4 telescope has a focal length of 600 mm and an image-scale of 344 arcseconds per millimetre. A 150 mm f/8 telescope has twice the focal length and an image scale of 172 arcseconds per millimetre.

How much of this image is presented to the eye of the observer depends on the size of the field stop aperture of the eyepiece. Let us say that we are using one of the standard $1\frac{1}{4}$-inch (31.7 mm) barrel-diameter eyepieces. Further suppose that it has a field stop aperture of 28 mm (it obviously has to be at least a little less than the size of the barrel). What field of view will we see when we look through it? Plugged into the 150 mm f/4 telescope we will see a field of view amounting to 28 × 344 arcseconds, or 9632 arcseconds. This is a field of view of 2°.68, or about 5 times the diameter of the full Moon.

With the 150 mm f/8 telescope the diameter of the field of view we can see is halved (though, of course, the image we see will be magnified twice as much). Our field of view now only amounts to about $2\frac{1}{2}$ times thediameter of the full Moon. That is one quarter of the area it was previously.

A field stop aperture of 28 mm is just about as large as we can get in any eyepiece assembly that has to fit into a standard $1\frac{1}{4}$-inch barrel; so you see the advantage of using the telescope of low focal ratio when we wish to image large fields of view – and the importance of choosing the correct eyepiece.

What design makes for a good rich-field telescope? In years past, short-focus refractors were traditionally chosen. They might have a two-element achromatic objective lens of about 130 mm aperture and a focal ratio of f/5 or f/6. However, refractors of this aperture were, and still are, very expensive and the problems with uncorrected colour, properly known as the *secondary spectrum*, precludes using them with even moderately high magnifications. So, these telescope were good for imaging wide fields of view, perhaps three or four degrees in extent, but were totally useless for anything else, such as Moon or planet observing.

Nowadays, improvements in optical glasses have enabled two-element objectives of considerably better colour correction to be manufactured, or even three element objectives of marvellous colour correction with similarly low focal ratios. The modern two-element, *semi-apochromatic* objectives (they are often marketed, though, as *apochromatic* though this term should be reserved for the very best objectives – normally needing three optical elements to give this performance) make fine rich-field telescopes and yet they also give acceptable performance as telescopes to use for planetary observation. Telescopes, even the low-focal ratio ones, fitted with the three element, truly apochromatic, objectives are superb for nearly all applications. They are, though, expensive.

One can have a reflecting telescope of the same aperture and the same (or maybe even smaller) focal ratio as the refractor but for a **very** much smaller outlay. Another advantage of reflecting optics is that they do not produce any chromatic aberration. The eyepiece which, of course, uses lenses will only give significant trouble in this respect if the inappropriate eyepiece design is chosen – see later for more on this.

The simplest, cheapest, and overall the most efficient design you could choose to use as a rich-field telescope is the Newtonian reflector. As ever, an aperture of 100–200 mm and a focal ratio of about f/4 to f/6 is most appropriate. They do, though, have a couple of flaws not possessed by the rich-field refractor.

One problem arises from the on-axis obstruction in the light path due to the diagonal mirror. Look again at Figure 3.1. The exit pupil has silhouetted within it the obstruction due to the secondary mirror of the telescope (in this case a 0.46m Newtonian reflector, with a secondary obstruction spanning a quarter of the diameter of the primary mirror). Normally, the observer is not aware of the secondary's silhouette but when the exit pupil is large enough, the silhouette can then be a substantial portion of the size of the observer's eye pupil. Then it does make itself felt.

Of course, using a magnification low enough to produce an over-large exit pupil is wasteful but one can still do so with the unobstructed refractor without incurring any extra penalties. To do so with the reflector is yet more wasteful of the light (only an annulus of the celestial light passing into the observer's eye) and can even produce an unpleasant shadowing effect if the silhouette is big enough. So, if you use a Newtonian reflector you really are limited to magnifications not much lower than the so-called "visual equalisation" value.

The other problem with the Newtonian reflector is that the usual paraboloidal primary mirror is afflicted with a particular optical aberration called *coma*. If the mirror is a good one it will produce good images close to the centre of the field of view. Stars, for example, will look as they should do – intense points of light – near the centre of the field of view. Stars away from the centre will be stretched out into small comet-like shapes (with the tails fanning outwards), the aberration increasing with distance from the centre of the field of view.

The field of view of acceptable quality, expressed in degrees, that a paraboloidal primary mirror can produce is inversely proportional to its diameter. The following relation can be used as a guide to the maximum diameter, L, of field of acceptable definition for visual observing at low magnifications:

$$L = 400/D,$$

where D is the diameter (or clear aperture) of the primary mirror in millimetres. Thus our 200 mm Newtonian reflector can potentially image a field of view of 2 degrees diameter with the star images seen at the edge of the field of view being only slightly distorted. Refractors are also troubled by outfield aberrations, though usually less so than is the case for the Newtonian reflector.

Fortunately, the limited coma-free fields of Newtonian reflectors can be expanded by means of correcting lenses plugged in just before the eyepiece (in the same manner a Barlow lens is normally used).

An example is TeleVue's "Paracorr" lens, though this unit does produce a 1.15× increase in the magnification, and so reduces the field of view an eyepiece gives to 0.87× the value without the corrector. That may still be preferable when the stars are sharp right to the edge of the field, rather than cometic as before!

On the subject of "plug-ins", I should mention that a number of companies market telecompressor lenses. They will effectively reduce the focal ratio of a telescope but usually at the expense of some *vignetting* (darkening of the outer parts of the field of view) and an increase in the outfield aberrations – so you don't get the increase in the usable field of view you might expect. If you do go in for one of these, make sure that it is one intended for visual use. Most are only suitable for photography or CCD imaging.

At the time I write these words, *Meade* have just brought on to the market a series of Schmidt-Newtonian telescopes: a 152 mm f/5; a 203 mm f/4; and a 254 mm f/4. Even the 254mm telescope costs less than $1K in the US. The adverts state that the coma afflicting them is half the extent of the equivalent Newtonians. As such they should make excellent rich field telescopes but do bear in mind that their low focal ratios demand expensive eyepieces if they are to provide quality imaging.

When it comes to the mechanics of mounting the telescope, a simple altazimuth mounting, such as a Dobsonian, will suffice for any telescope to be used solely for low-power visual observing. Indeed, it may even prove more convenient than an equatorial for such tasks as sweeping for new comets. As always, smooth motions and a vibration-free mount are the primary concerns.

While I have extolled the virtues of the rich-field telescope in cometary work, I must also make it clear that almost any size and type of telescope can produce worthwhile results. Most of today's amateur astronomers purchase Schmidt–Cassegrain telescopes to carry out general visual observing and CCD imaging, for instance. Whatever equipment you possess, please do press it into service for observing comets. When purchasing equipment spend your money wisely – and when using your equipment, choose the observational tasks most suited to its design.

Choosing the Best Eyepieces for Your Comet Observing

It is worth taking some care when one is choosing a particular telescope to purchase, given the almost bewildering array available nowadays. This is even more the case when choosing the eyepieces to use with it. The modern generation of eyepieces can swallow up a sizeable chunk of the budget for the whole installation. Few appreciate until it is too late that choosing the correct eyepiece(s) for the observational task is just as important as choosing the most suitable telescope.

The references given earlier cover all general matters concerning eyepieces as well as dealing with specific designs and their pros and cons.

One important numerical relationship that is worth giving here is:

$$\text{real field} = \text{apparent field} \;/\; \text{magnification}$$

In using this equation, the values of real and apparent field must be in the same units. Most usually, they are both expressed in degrees.

For all practical purposes the *apparent field* of the eyepiece can be defined as the angle through which the observer's line of sight must swivel in order to see from one extreme edge of the field of view to the other. This quantity is a characteristic determined by the design of the eyepiece.

As you might be aware, apparent fields of view of more than 55° come from complex designs of five to eight element eyepieces – such as the TeleVue company's 'Radian' (60°), 'Panoptic' (68°), and 'Nagler'

(82°), and Meade's "Super Wide Angle" (67°), and "Ultra Wide Angle" (84°). These cost hundreds of dollars each, so you may well have to set your sights lower. Plössyl eyepieces are very easily available from a number of manufacturers (mostly in four-element designs, sometimes five element) and they usually have apparent fields of view in the range of 50° to 55° and cost circa $80 at current (2001) prices.

The *real field* factor in the equation is the angular extent of the sky that the observer actually see when he/she uses a particular eyepiece with a particular telescope. The diameter of the full Moon subtends just over $\frac{1}{2}^{\circ}$, for instance, so you will need a real field of at least this size in order to see all the full Moon in one go. As an example, if an eyepiece of 60° apparent field of view gives you a magnification of ×40 when it is plugged into your telescope you will see a real field of 1°.5 (almost three Moon-diameters).

One thing I should warn you about is that you may well get a slightly smaller real field in practice than you might expect using the equation with the manufacturers stated value of apparent field for the eyepiece. Almost all eyepieces suffer to a small degree from pincushion distortion. That is, the magnification of the image increases a little away from the centre of the field of view. This aberration is usually worst in eyepieces of large apparent field. For instance, a Nagler eyepiece, with its stated "82°" apparent field, will behave like an eyepiece with an apparent field closer to 78°, as far as the equation predicting the real field is concerned.

The effects of chromatic aberration and the various seidal aberrations (spherical aberration, coma, astigmatism, curvature of field and distortion) will make themselves felt if you use an eyepiece of too-simple a design on a low focal ratio telescope. Achromatic Ramsdens and Kellners are just about acceptable for low and moderate power views with telescopes of focal ratio not much lower than f/6. I say this in the face of the fact that manufacturers often supply Kellner eyepieces with f/4.5 Dobsonian telescopes! They have fields of view of around 40°. Their one virtue is that they are cheap.

A Plössyl eyepiece is only just about good enough for very low power views (the eyepiece having a focal length of not much less than 25 mm) on a telescope of focal ratio around f/4. The situation is much better at f/5, the Plössyl then performing well at low and moderate powers. At f/6 it will deliver good images over the full range of magnifications. The best Orthoscopic eyepieces can out-do the performance of the Plössyl type, though the apparent field of view is a little smaller, being in the range of 40–45°.

Most of the more complex designs will work well with low focal ratio telescopes, though the quality of the images near the edge of the field of view will suffer at the very lowest ratios even with these eyepieces. The expensive, but phenomenal, Nagler eyepieces are the champions at focal ratios as low as f/4.

When you are selecting equipment for your observing, I recommend that you consider the eyepieces in combination with the telescope, rather than treating them as an afterthought. In particular, consider the cost of the eyepieces you will need when selecting the focal ratio of the telescope. You will be able to achieve the biggest field of view at the smallest cost if you balance the focal ratio against the eyepiece type needed.

Binoculars for Observing Comets

Much has been made about binoculars being superior to very small telescopes. In most cases I would agree. It all depends what one intends doing with them. There is one oft-given piece of advice, though, that I would always dispute. Many authorities recommend 7 × 50 binoculars (magnification factor 7, 50 mm diameter object glasses) as the best size you can choose. However, this produces an exit pupil of 7 mm diameter at each of the eyepieces. Fine if you are in your teens but wasteful if you are much older than that. For example, if your eye pupils will only expand to 5 mm when fully dark adapted, then you might just as well have chosen the equally common 7 × 35 binoculars. Your views through either will be indistinguishable and the 7 × 35 binoculars will be much cheaper, as well as more compact and lighter to carry and handle.

The binoculars I recommend as best are the 10 × 50 size. Their 5 mm exit pupils will suit the eyes of people aged from 6 to 60. Remember also our earlier discussions about the best magnifications to use to see stars and nebulous, comet-like, objects: the very lowest magnification is certainly not always the best one to use.

What about higher magnifications, though, and perhaps binoculars with bigger objective lenses? One advantage of binoculars over telescopes is their inherent wide field of view. A typical mid-priced binocular will have eyepieces fitted to it that have apparent field diameters of about 50°, or maybe just a little larger. A magnification of ×10 will result in a real field diameter of 5°. That is, for instance, just enough to squeeze all the main stars of the Hyades star cluster into the field of view in one go. Binoculars in the same price-range giving higher magnifications necessarily have smaller fields of view. The prices of binoculars fitted with eyepieces of much wider apparent fields really starts to rocket upwards.

There is another disadvantage to binoculars of higher magnification. You will find that holding them steady enough is not at all easy. A few moments of trying to fix your eye on the subject while it jitters and swims erratically about in your vision will make you feel queasy. Not only that, your eye and brain will have precious little opportunity to actually study the subject. You will find you can actually perceive remarkably little in the way of detail, despite your best efforts. You might wish you had chosen the lower power binoculars (and saved some money), instead!

Many comet observers use 80 mm binoculars but what about binoculars with even larger object-glasses? Yes, the increased light grasp is usually advantageous, though the eye pupil diameter is always the limiting factor in the lowest magnification (and so the widest field of view) we can use without wasting light.

Even more pressing – please excuse the pun – are considerations of weight. How long will you be able to hold a hefty pair of binoculars in position before the ache in your arms and shoulders, soon spreading to your neck, gets too much to ignore? Not very long. All-in-all you will find 10 × 50 to be the best category of hand-held binoculars for your comet observing.

Even with 10 × 50 binoculars, you will soon see the advantage of resting your elbows on a suitable platform when using them for long periods. The top ledge of a fence, or the platform of a bird-table or anything else you can press into service will increase your comfort – and hence the detail you will be able to appreciate and observe – when using such a rest to ease your muscle load.

Of course, the situation is a little different if the binoculars themselves are on some sort of stand (see

Figure 3.5. The Orion 80 mm Binocular Telescope. Courtesy Orion Telescopes and Binoculars.

Figure 3.5). Then they can be heavier and even have more magnification, and still be a joy to use. As is the case when you are choosing telescope mountings – beware anything too light and spindly. A number of different mounting are commercially available.

There are also now available other mountings for binoculars that work on the same general principle as the table-lamps of the famous "Anglepoise" tradename. Some have counterweights, others have springs to do the same job. In some cases you can attach these units to the arms or frames of deck-chairs, making for very comfortable viewing! I recommend you scan the adverts if one of these appeals to you.

Of course, big mounted binoculars are extremely expensive. For instance, Fujinon's 25 × 150 binoculars come on a sturdy tripod and tilt-and-pan head. They also have extra prisms incorporated in the optical train in order to have the eyepieces angled upwards by 45°. There are some others angled upwards to 90°. This feature affords a great relief on the neck muscles when viewing any object at high altitude. They are well-liked by Japan's most successful comet discoverers. It is a shame that a new pair will cost you over $10 000!

Binoculars, large and small, can be found second-hand but please do examine them closely (particularly looking for scratches in the lenses and checking that the focusing action is smooth and slop-free) and thoroughly check them out in use before you part with any money. The same basic advice goes for any optical device, large or small, that you intend to buy: buyer beware!

Image stabilised binoculars are of particular interest because they give a very steady view at higher magnifications without the need for a tripod. The effect of the image stabilisation technology when hand-holding higher power binoculars is amazing. When switched off objects jiggle around in the field of view just as they would in conventional binoculars. Press the stabilisation button, however, and everything becomes rock solid. Observers report a significant improvement in visual limiting magnitudes and perceived details over non-stabilised binoculars of the same aperture. The downside is that image stabilised binoculars are very expensive. The cheapest Canon stabilised binocular, the 12 × 36 IS, retails for around $800 in the US at late 2001 prices. The much more capable 18 × 50 IS model costs around twice as much. This may sound a lot but if you get to use one of these instruments under a dark, clear, sky you will understand why binocular observers are so enthusiastic about this new technology.

So much for our all-too-brief survey of the most important issues in connection with the selection of the equipment that would be most suitable for your comet observing. In the next chapter we get down to the subject of the actual observing.

Chapter 4
Visual Practices

Amateur astronomy has undergone a revolution in the last few years, thanks to the advent of affordable sophisticated technology. A modern amateur's computer-controlled telescope can find and then track any celestial body. The CCD astrocamera plugged into it can obtain images that years ago would have required a large professional telescope. The amateur's computer can perform image processing, and even astrometry and photometry. Tasks that were the sole province of the professional astronomers of not that many years ago can now be undertaken by amateurs. As you might expect, the main bulk of this book reflects this new astronomy.

However, do not despair if you cannot afford equipment such as this. Even if you can, it might be that computers and technically complicated telescopes are not to your taste. There is much pleasure to be gained by using "eyeball" and "manual" methods. Furthermore, not all of the older methods are obsolete. There still remains the opportunity for making a genuinely useful contribution without going anywhere near a computer, or a computer-controlled telescope. In this chapter we will concentrate solely on "low-tech" visual methods for observing comets.

First Locate Your Comet

Let us begin with the premise that you know that there is a comet visible somewhere in the sky and you wish to observe it. Only the brightest comets ever make the general press, so you probably knew about the comet from belonging to a local and/or national astronomical society (via "circulars/newsletters", published ephemerides, "journals", or personal communication), or you were alerted to it by a magazine article, or found out about it on one of the Internet web sites.

When the recent great comets Hyakutake and Hale–Bopp were near perihelion, they were prominent enough not to need any particular aids in finding them – a directive such as: "low in the western sky at mid-evening" was sufficient. Those comets, though, were exceptional. Away from perihelion even they appeared small and faint, of course.

Normally you will need some form of assistance in order to track down any particular comet. What will give you the best help will depend on the apparent brightness of the comet and the particular instrumentation you intend to view it through. I recommend you use the quickest and simplest aids allowable, given the brightness of the comet.

Locating Comets Visible by Eye or Through Binoculars

Where reasonably bright comets (needing no more than binoculars to see them) are concerned the various glossy magazines are useful in that they often give partial sky maps on which the comet is plotted for specific dates. If you know the sky at least reasonably well that should be enough for you to track down the comet.

Your telescope might have good setting circles. That could be sufficient if you have the figures from a published ephemeris which you can use. Even then, you might still have to "stir the telescope about a bit" in order to find the comet if the field of view of the eyepiece is not very wide.

Failing that, you might use one of the many available general star atlases in conjunction with the chart of the small area given in the magazine. A few moments indoors locating the area covered by the chart should enable you to locate the specific area on the atlas where the comet is expected.

In slightly more difficult cases you could take your atlas and the magazine, and your red-filtered or at least dim (to avoid ruining your dark adaption) torch outside and find the general part of the sky by means of the pattern of brightest stars on the atlas, and then the dimmer ones, to narrow down the area of search to the area covered by the magazine map.

At this point you can bring your binoculars into play. Once you locate the comet through the binoculars you can then aim your telescope, which of course should be fitted with its widest-field eyepiece. Once again looking for recognizable patterns of stars (taking into account the larger magnification – and the fact that the telescope produces an inverted field, unlike the erect view through the binoculars), you should soon locate the comet.

Failing binoculars, perhaps you can use the finder-telescope. However, this will usually be more awkward and, anyway, many finder telescopes are too small to be of much use in locating "faint fuzzies". Everything is eased considerably if the main telescope is fitted with an eyepiece of at least 2° field of view.

A more systematic approach is to use *star-hopping*. More about that and how to find comets too faint for binoculars shortly. But first ...

Measuring the Real Fields of Your Eyepieces

It is extremely useful to know the values of real fields of view that your eyepieces produce with your telescope. You can get a good idea using the equation that relates real field, apparent field and magnification (see page 59) but that does rely on the manufacturer's stated values being accurate. You can obtain a value which you know to be accurate by means of a quick and very easy method. The only addition to your equipment you will need is any timepiece that you can use in the manner of a stopwatch.

With the eyepiece plugged in and focused, set your telescope on a known star. One near the celestial equator would be a good choice. If your telescope has a motor drive switch it off. If your telescope has a properly aligned equatorial mounting (though a polar-pointing inaccuracy of a degree or two will not matter for this) simply set the star close to the centre of the field of view. With the declination axis firmly clamped, release the RA clamp and move the telescope so that the star moves eastwards across the field of view and just disappears beyond the edge of it. Apply the RA clamp and get ready with your stopwatch.

Start the stopwatch as soon as the star appears and stop it as soon as the star has disappeared at the western edge of the field of view. Repeat this a number of times and take an average of the time, t, the star takes to track across the full diameter of the field of view.

The procedure is virtually the same if your telescope is altazimuthly mounted, except that you must ensure that the star passes through the centre of the field of view. If not, then you will be timing the star across a chord through the eyepiece field rather than the full diameter of it. You might need to do a couple of preliminary trials to judge where to place the star along the eastern edge in order that it does cross through the centre of the field of view. Reject all the timings where it obviously "misses the target".

At the celestial equator the stars all move along at 15 arcseconds per second (the difference between the sidereal and the solar day producing an discrepancy far too small to bother with here). This is a quarter of a degree per minute. Consequently, if the star you selected lies on, on very close to, the celestial equator then the following equation will convert your timing to a value of the field of view of your eyepiece:

$$D = 15t,$$

where t is the time in seconds and D is the diameter of the field of view of the eyepiece, measured in arcseconds. Divide this number by 60 and you have the real field of view of your eyepiece expressed in arcminutes. Divide by 60 again and you have the real field in degrees.

Stars that are away from the celestial equator appear to move more slowly. In that case a factor $\cos\delta$ is needed, where δ is the declination of the star. You can estimate δ from your star atlas, or find it from a catalogue, etc. The equation to use is then modified to:

$$D = 15t\cos\delta$$

What are the fields of view of all of your eyepieces? Knowing the field of view of at least your widest field eyepiece, will help you use a more sophisticated method for visually finding your prey. This method really comes into its own for finding faint comets.

Locating Faint Comets – Preparing a Finder Chart and "Star-hopping"

First select a good star atlas. One I can recommend is: *Uranometria 2000.0* by Wil Tirion with Barry Rappaport and George Lovi. Nearly a third of a million stars are plotted, the faintest being of the tenth magnitude. The scale of the charts averages 0°.6 per centimetre and there is a convenient grid superposed. Smaller, and less expensive, atlases can also be pressed into service. Another atlas by Tirion is: *Sky Atlas 2000.0*. It plots the positions of 43 000 stars to eighth magnitude on charts with a scale averaging 1°.2 per centimetre. If you have a computer you can also print-out charts from one of any number of available planetarium programs. More about those in Chapter 6.

So, we have our chart. Next we locate the position of the comet we wish to find. To avoid marking the atlas, I recommend photocopying the appropriate part of it. Of course, there is no need if the chart is a computer print-out.

Then note the scale of the chart in the region of interest. The scale will vary a little across the chart but measuring the grid lines or abscissas will soon enable you to make a good estimate of the scale at the point where the comet is. Next draw a circle, centred on the position of the comet, **with a radius that corresponds to the radius of the field of view of the locating eyepiece.**

An alternative to photocopying the atlas is to use a transparent acetate sheet and use a fine-tipped spirit marker to draw the circle on this. Use the dot at the centre of the circle to position the sheet over the atlas at the predicted position of the comet.

Divide the circle you have drawn into a series of squares, as shown in Figure 4.1(a). On another piece of paper draw yourself a large circle to represent the field of view of the locating eyepiece. Divide this one up into identical squares. Then draw in the corresponding positions of the stars as they appear in the photocopy, or are framed by the grid on the acetate sheet. There is no need for painstaking precision. Quickly made eye estimates will do fine. The purpose of the grid is to help you do this. You will wind up with something like that shown in Figure 4.1(b). This is your finder-chart.

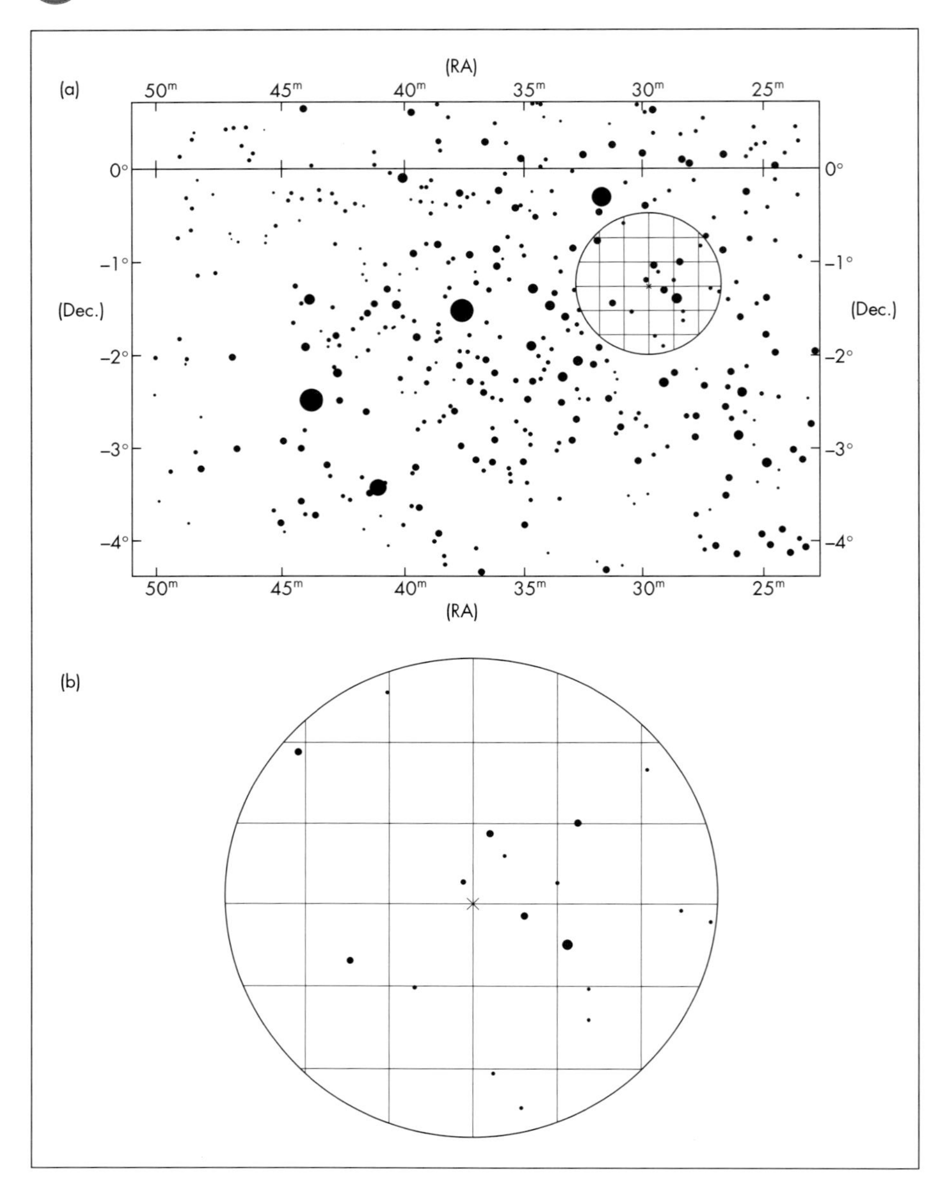

It should only take you a few minutes to create a finder-chart and, by helping you recognise the star patterns in the field of view, it will save you a considerable amount of time an effort at the telescope eyepiece.

Figure 4.1. The preparation of a finder-chart. Details in text. In this case an eyepiece real field of 1°.5 is represented centred on coordinates $\alpha = 05^h\ 29^m.8$ and $\delta = -1°$.

Using an overlay sheet or a photocopy is also useful for "star-hopping". Draw a set of nearly (or slightly) overlapping circles, each representing the telescope field of view, that links an easy-to-find object (such as a bright star or recognisable asterism) with the position of the comet. This gives you a set of "visual stepping-stones"; you move the telescope until the next pattern of stars enclosed by the next circle comes into view, etc. After several of these controlled "hops" you arrive at the field of the comet.

Two or three minutes spent "star-hopping" is certainly preferable to half an hour of stirring the telescope about hoping that the comet will swim into view. Even if it does, it might be too faint for you to detect while it is lurching across the eyepiece field!

In the most difficult cases you might want to prepare two finder charts: one for your finder-scope, and one for the main telescope. Do also bear in mind that comets can stray a little from their predicted paths. This is especially the case for newly discovered ones, before their orbits have been accurately established.

Second only to using modern technology, a finder chart will give you the best chance of seeing the predicted return of a periodic comet, or it will allow you to see any newly-discovered comet, at the earliest possible stage. Then you will have the opportunity of producing a long and really useful series of observations as you follow it through perihelion passage and beyond.

Observing and Drawing Comets

You have found your comet. Now to observe it. As I discussed at length in the last chapter, do use a range of magnifications to study the various features of the comet.

Let us break down the comet into two parts – the details visible close to the false nucleus and the comet as a whole – and consider each in turn.

The Immediate Surrounds of the False Nucleus

This brightest region of the comet is especially interesting. Within it lurks details very often lost to "white-out" in photographs or CCD images which are exposed to render the fainter parts of the comet. Yet here dramatic physical processes occur that define the very "beating heart" of the comet.

Look carefully at the false nucleus. How bright is it? Try the full range of magnifications you have at your disposal. You will often find that this part of the comet can stand very high magnifications – and you might see details revealed that would remain indistinguishable at lower powers. The false nucleus is usually very small, almost star-like. However, in very active comets it can expand in size to several arcseconds across. What is its shape? Is it spherical, or elongated, or even irregular? A high magnification is a must, here.

Are there any jets visible issuing from the false nucleus? They usually emerge pointing in a sunward direction for a few arcseconds, or maybe arcminutes, and then curve round to trail off in the direction of the comet's tail. What about hoods, shells, jets and rays?

Provided they are carefully executed, drawings based on what you see through the eyepiece still have real scientific value. This is especially the case for drawings of these innermost part of the comet. For instance, you or the co-ordinator to whom you send your observations, may have a chance to determine the rotation period of the nucleus from a sequence of your (or yours and other people's) drawings. This way of determining the rotation period can give more definite results than by photometric (brightness) measurements.

One spin-off (no pun intended) is the possibility of determining the inclination of the rotation axis of the comet. This demands that the orientation of your drawing is carefully determined and marked. Another spin-off is that the regions of particular activity (the roots of the jets) on the nucleus could be mapped. What is possible depends, though, on the details visible (which you cannot control, beyond using the best possible technique) and the quality of your recording/drawing (which is your responsibility).

Building up sequences in the course of a night, and from night to night, are particularly instructive. Look at the drawings displayed in Figures 4.2 to 4.4 and you will see how much detail can be recorded by means of eye and pencil. You will find other examples of drawings in Chapter 2 (Figures 2.2 and 2.3).

Increase the ease, and hence the accuracy, of your drawings by making yourself some form of clip-board to hold your drawing paper. Fit a dim light at it's head. Arrange a shield over the bulb so that your eye cannot see the bulb itself. You only want to see the paper illuminated. A red filter over the bulb will help preserve your dark-adaption. Even better if you include a potentiometer to vary the brightness. A terry-clip can be used to hold a battery onto the board. Keep it simple, though, and above-all keep it lightweight.

As for the drawings themselves, well, pencil and paper or any other combination of artistic media can be used. Black on a white background is fine, though you might prefer white on black. Everyone will have their

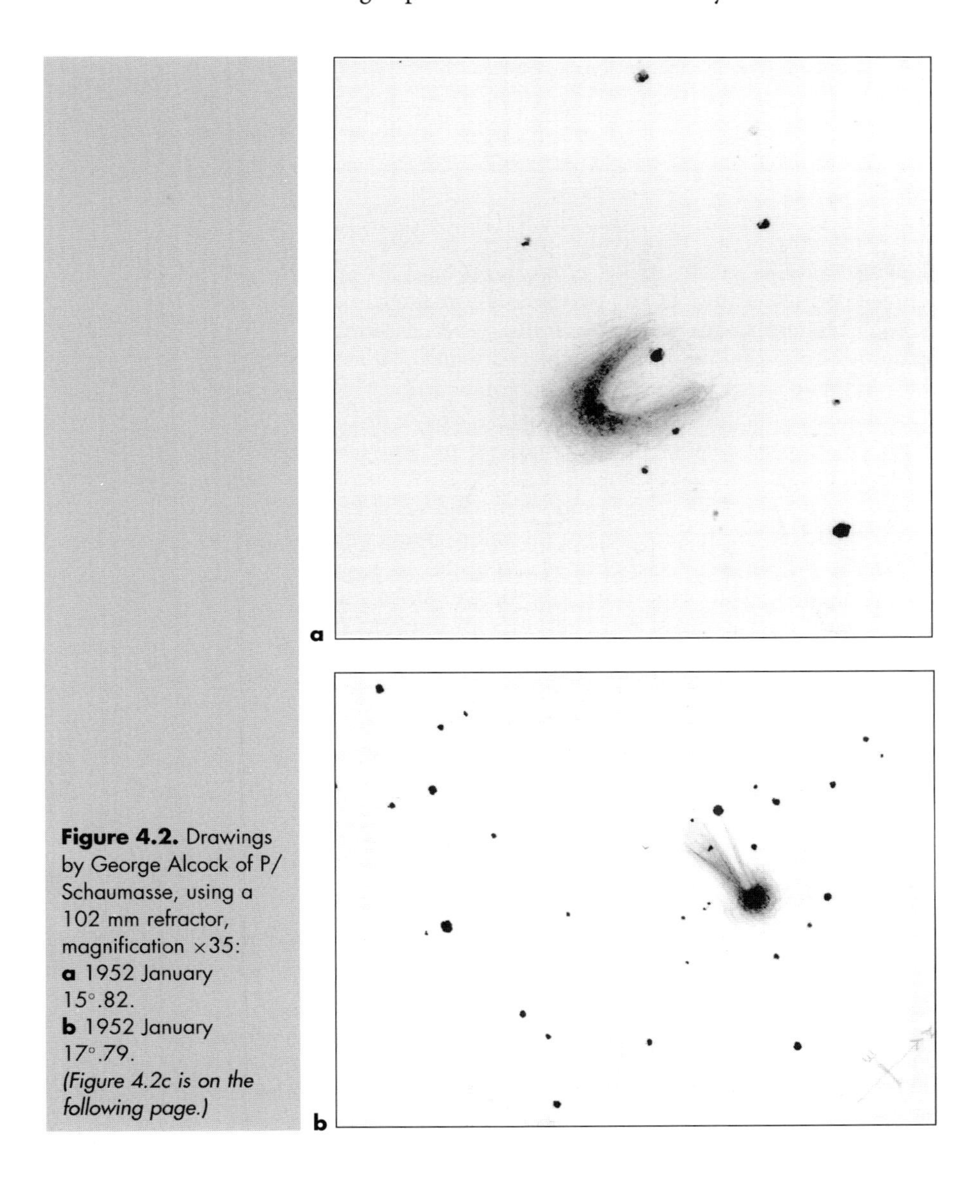

Figure 4.2. Drawings by George Alcock of P/ Schaumasse, using a 102 mm refractor, magnification ×35: **a** 1952 January 15°.82. **b** 1952 January 17°.79. *(Figure 4.2c is on the following page.)*

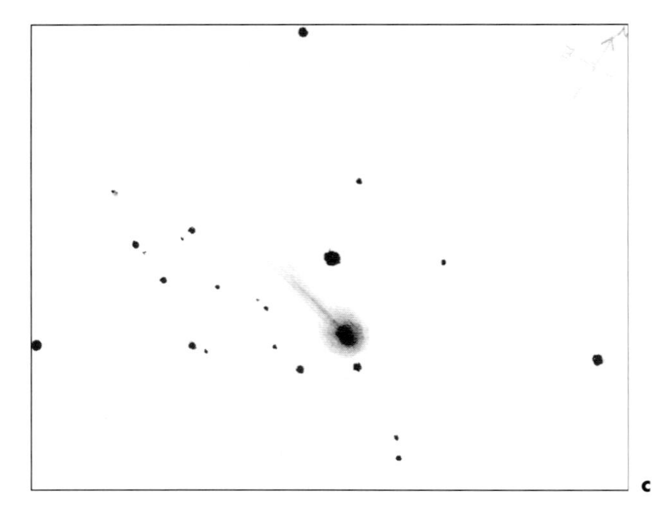

Figure 4.2. c 1952 January 19°.79.

own preferences; all that really matters is that the drawing is as accurate in its proportions and in its represented intensities as you can possibly make it.

You can brush-up your observing and drawing techniques on suitable deep-sky objects, while you are waiting for the next comet to appear. You will find that considerable practice is necessary before you will be happy with your results but the end result really is worth it. So, go on, have a go!

Take special care with all the proportions and sizes. Try to accurately record the positions of the field stars, as well as the comet – in fact I recommend doing this as your first main task. The stars, once placed, will help you achieve positional accuracy in the rest of your drawing. If you find that sketching a faint grid of lines on the drawing paper helps you - then do it. You will have your own way of working but, I say again, the most important aspect of your final drawing has to be its accuracy.

The Overall Appearance of the Comet

A comet's coma rarely appears perfectly circular. Its shape and size, its brightness and the distribution of brightness within it reflect the vigour and type of activity of the nucleus, as well as the character and extent of its interaction with the solar wind. This is even more true when one considers the comet's tail or tails. The photographs, drawings and CCD images presented in this book, as well as the much larger number presented in the accompanying CD-ROM, illustrate the enormous diversity of the sizes and forms comets can take.

Even more confusing is the fact that some comets can display significant brightness and structural changes over

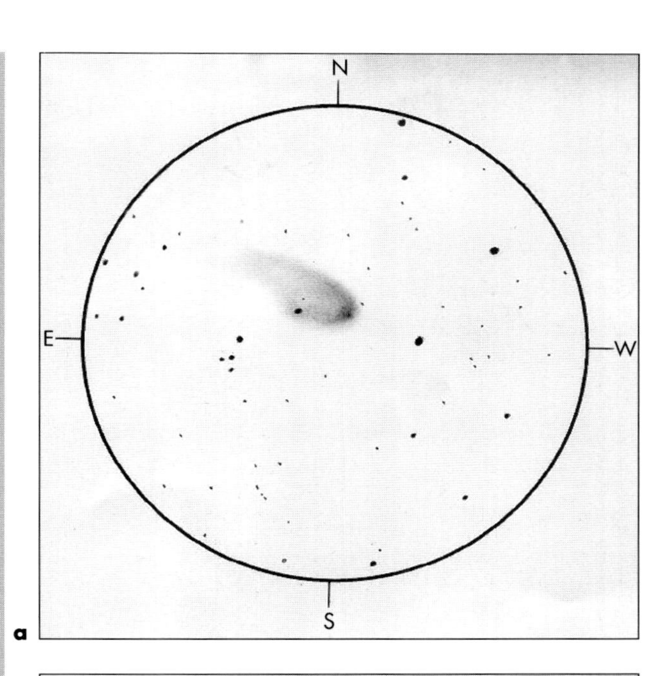

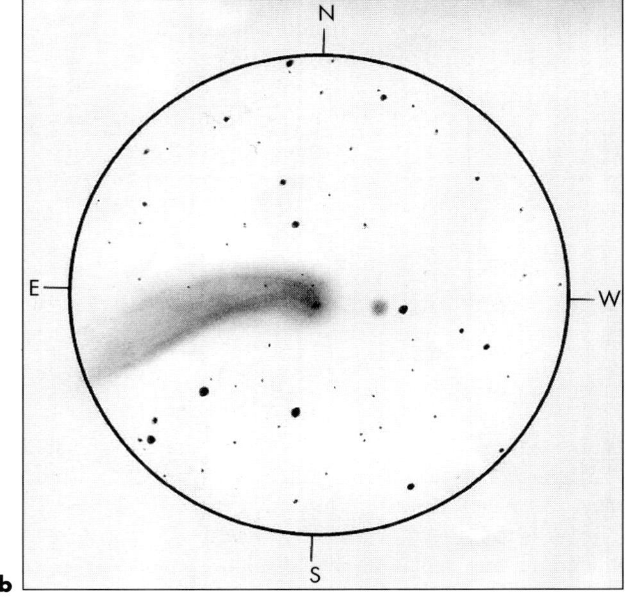

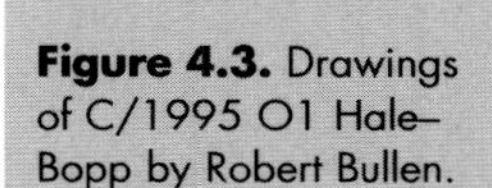

Figure 4.3. Drawings of C/1995 O1 Hale–Bopp by Robert Bullen.

a
Date: 15th September 1996
OBS: 16
Log: 10
GMT: 20:50–21:20
Cond: III
Sky: III
RA: 17h 31.6m
DEC: –5°.42′
11 × 80 binoculars
Field size: 4.5 deg
Mag: 5.1

b
Date: 1st October 1996
OBS: 19
Log: 12
GMT: 19:00–21:00
Cond: II
Sky: II
RA: 17h 29.7m
DEC: –5°.03′
11 × 80 binoculars
Field size: 4.5 deg
Mag: 4.7
OBS made in the Spanish Pyrenees

(Figure 4.3c, d and e are on the following page.)

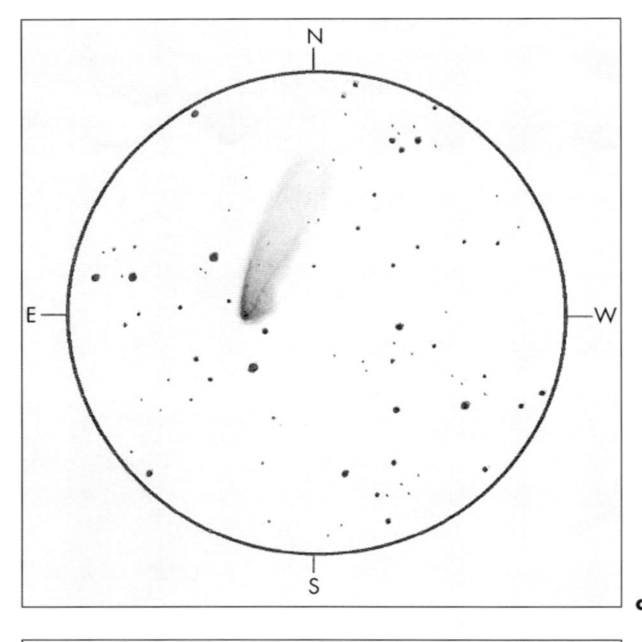

c

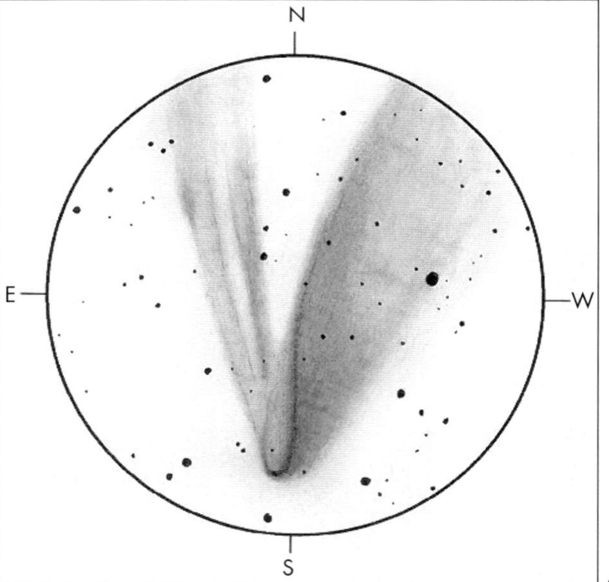

d

Figure 4.3c and d.

c
Date: 25th January 1997
OBS: 28
Log: 15
GMT: 05:00–06:00
Cond: III
Sky: III + 16 day old moon
RA: 19h 25.46
DEC: + 12°.27′
11 × 80 binoculars
Field size: 4.5 deg
Mag: 2.5

d
Date: 28th March 1997
OBS: 53
Log: 32
GMT: 20:00–21:45
Cond: III
Sky: III > IV by 21:15. Some low and high level cloud interfered with the OBS at times
RA: 1h 17.0m
DEC: + 45°.23′
11 × 80 binoculars
Field size: 4.5 deg
Mag: –1.3

a period of less than an hour. All comets show considerable changes over much longer periods (hours, days or weeks). How does one make sense of all this? The answer is to break down the various aspects or features of the comet and separately try to quantify them.

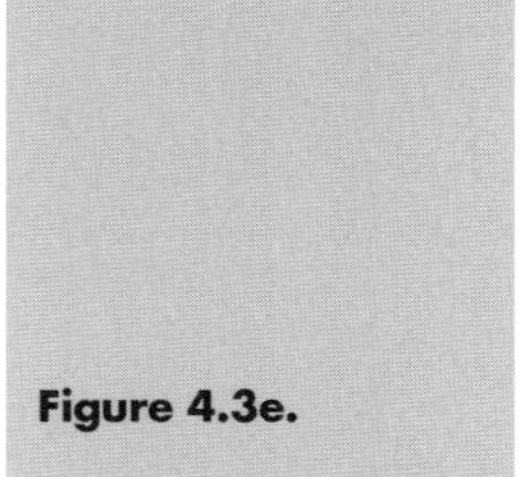

Figure 4.3e.

e
Date: 5th April 1997
OBS: 57
Log: 39
GMT: 20:00–22:00
Cond: III
Sky: III
RA: 2h 29.6m
DEC: +42°.09′
11 × 80 binoculars
Field size: 4.5 deg
Mag: –0.8

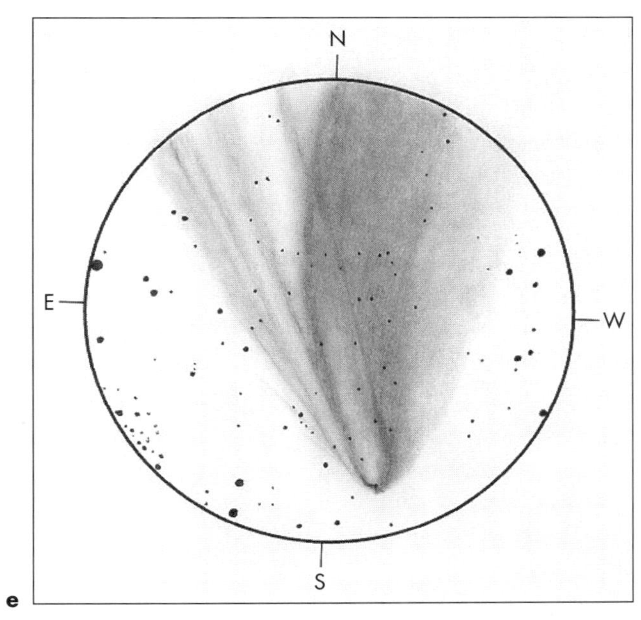

e

Figure 4.4. Drawings by Dr Richard McKim showing the development of the tail of C/1996 B2 Hyakutake.

To begin with, here are a few pieces of data you should attempt to determine whenever you visually observe a comet:

- How condensed is the false nucleus and coma? There is a scale for this. The quantity is called the *Degree of*

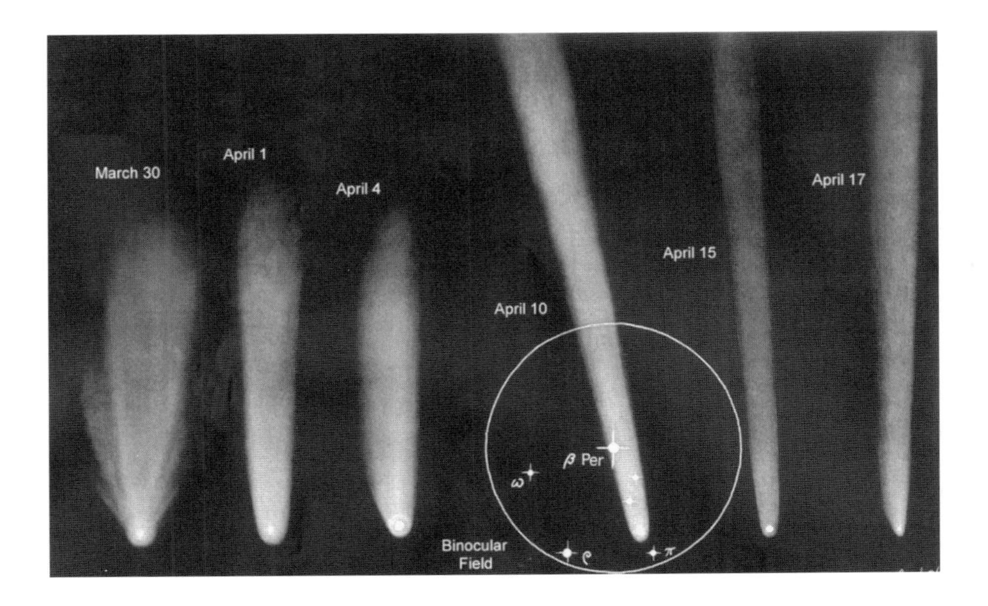

Condensation (*DC*). The scale runs from 0 to 9. A comet with a DC of 0 appears totally diffuse while one with a DC of 9 appears star-like. You should use the "key" of comet appearances and associated DC numbers presented in Figure 4.5. as an aid to making your determinations.

If there is an additional feature visible in the coma, then add "D" after the DC value if this feature is disk-like and makes a significant contribution to the total brightness of the coma and changes the DC value by more than 2. If the disk-like condensation makes a less significant contribution to the overall brightness than register its presence by adding a "d" after the DC number. If the condensation is star-like then add an "S" or an "s", depending on whether it changes the overall DC number by more or less than 2, respectively. If any condensation is intermediate between being disk-like or star-like then add an "N" or an "n" if it is bright or faint, as for the other added letters. You will find a couple of representations of a cometary coma with condensations in the lower-right two cells of Figure 4.5. In these hypothetical cases the DC numbers assigned to them would be 2S and 3D.

Of course, the best you can do is to make an estimate and it is bound to be an imprecise one. Further, the value of DC you determine will often differ when you view the comet is viewed through different equipment! Nonetheless, the DC value is still much used in comet studies.

- What shape is the coma? Is it circular, or some other symmetrical shape, or does it tend to appear fan-shaped or parabolic, or is it irregular in outline? Linked to the foregoing: how diffuse does it appear and what structure is visible within it. How is the brightness distributed? For instance, is the coma diffuse all round, or does it have a sharper edge along one side?
- How long is the tails (or tails)? What is (are) the shape of the tails(s) – fan shaped, narrow, curved or straight. Type 1 tail? Type 2 tail? How bright? How is the brightness distributed throughout the tail(s).
- What features, such as knots, waves, disconnections, are visible in the tail?
- What colours are visible anywhere in the comet? Normally any colours are pastel shades if they are visible at all (most comets appear a nebulous grey,

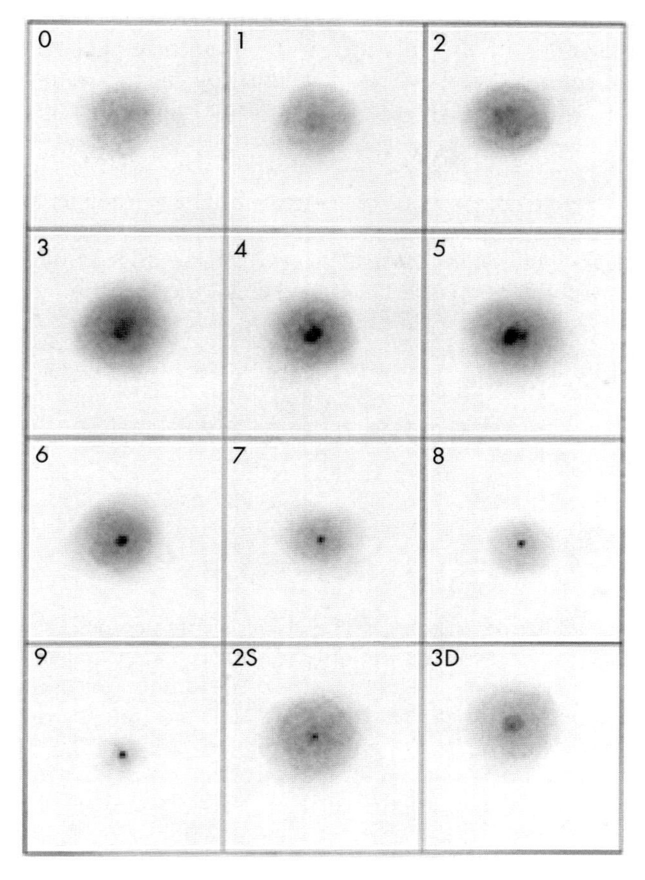

Figure 4.5. Degree of Condensation (DC). Note that the contrast of the images are here exaggerated for the sake of clarity. To assist in estimating a DC number, here follows a descriptive key:

0 Diffuse coma of uniform brightness
1 Diffuse coma with slight brightening towards centre
2 Diffuse coma with definite brightening towards centre
3 Centre of coma much brighter than edges, though still diffuse
4 Diffuse condensation at centre of coma
5 Condensation appears as a diffuse spot at centre of coma – described as moderately condensed
6 Condensation appears as a bright diffuse spot at centre of coma
7 Condensation appears like a star that cannot be focused – described as strongly condensed
8 Coma virtually invisible
9 Stellar or disc-like in appearance

(Diagram and key courtesy Jonathan Shanklin and the BAA)

even when seen through a telescope) but the really great comets can display very striking "plumages". The great Hyakutake of 1996 showed a greenish head and a prominent blue ion tail while Hale–Bopp of a year later was dominated by its glorious orange-yellow dust tail and coma. These colours were visible to the naked eye and were enhanced by viewing them through binoculars. The colours, especially those in the inner coma of each of these comets, added to the truly stunning spectacle when they were seen through the telescope eyepiece.

In addition to your drawings and simple notes there are some other measurements and other determinations you can make. These are briefly discussed in the following sections.

Brightness Estimates

To gauge the brightness of a comet's coma defocus the telescope/binoculars until the star images are expanded into disks approaching the size of the coma. Get them looking as similar as possible and then compare the intensities of the disks. The situation is easier if you have at least one star brighter and one star dimmer than the comet in the field of view. Then you can estimate the brightness of the coma as a fraction of the difference in brightness of the two stars. This is akin to the *Fractional Method* well known to variable star observers and was adapted for use with comets by G. Van Biesbroeck of the Yerkes Observatory in the second decade of the twentieth century. Variations in the techniques exist but the description given here is based on the *Bobrovnikoff method*. This is the most widely used and is certainly the easiest for the novice to master and get the most accurate magnitude estimates as a result. It does, though hinge on the comet being near stars of known **visual** magnitude ("photographic", i.e. blue, magnitudes will certainly not do). I recommend the Tycho Catalogue as a source of visual stellar magnitudes.

For instance, if star A is the brighter and the brightness of the coma lies three quarters of the way from the brightness of star A to star B (hence, closer in brightness to star B), you would write your observation down as: "A 3 V 1 B".

If star B has a magnitude of $8^m.4$ and star A has a magnitude of $7^m.2$, then the difference between them is $1^m.2$. One quarter of $1^m.2$ is $0^m.3$, so V is $0^m.3$ brighter than star B and is $0^m.9$ fainter than star A. Either way, the brightness of "V", the coma, comes out as $8^m.1$ (remember, dimmer stars have larger magnitudes).

In the absence of conveniently bracketing comparison stars, one has to estimate fractional "steps" in brightness and try to use comparison stars as close as possible in brightness to the coma. This is akin to the *Pogson Step Method* much used by variable star observers. Use as many comparison stars as possible and record the brightnesses in the following way (here the figures are hypothetical): "V = A + 4x", "V = B + 2x", "V = C + 1x", etc. and x represents the size of one brightness "step".

Use plus signs in the equations if the comparison stars are brighter (have more negative magnitude values) than the cometary coma, and negative signs if the comparison stars are dimmer than the coma.

Notice that V, the magnitude of the coma is common to each of the above equations. Therefore, taking just the first pair:

$$\mathrm{A} + 4x = \mathrm{B} + 2x,$$

so,

$$\mathrm{A} - \mathrm{B} = -2x,$$

or,

$$x = -\frac{1}{2}(\mathrm{A} - \mathrm{B})$$

using the foregoing hypothetical figures as an example. Putting in the values of magnitudes of stars A and B (looked up after the observation) gives us the value of x, the size of our "step". Repeat the same with all pairs of stars, B and C, etc., and so find an average value for x.

Lastly, put your average value of x into any one of the original equations (V = A + 4x, etc.) to find V, the magnitude value of the comet's coma.

Use this technique with different sized telescopes and you will get different magnitudes for the same comet! In general you should aim to use the smallest aperture that will show the comet well enough to make a determination. Of course, any large telescope can be turned into a small one by the simple addition of a cardboard mask over the aperture. Your 0.4m Newtonian telescope can be turned into a 100 mm one if the mask has a 100 mm hole in it (positioned off-axis in order to avoid the central obstruction)

Choosing a magnification not much higher than that required to produce a 5 mm exit pupil is desirable, though that might pose a problem if you are stopping down a large telescope (do you have an eyepiece of long enough focal length?). As ever, you will have to be prepared to make the best use of whatever equipment you have got. Do, though, be diligent in recording **all** the details of the equipment and methodology you use for the benefit of anyone who will use your observations.

The same general technique can be used for gauging the brightness of the false nucleus, less defocusing being necessary in that case.

Measuring Sizes and Position Angles

The most accurate visual method to find sizes for any object within the field of view is to use an eyepiece micrometer. However, few amateurs have these instruments nowadays and since space is limited here I will refer you to Chapter 14 of the book *Advanced Amateur Astronomy* (full reference given on page 44) for details of these instruments and how to use them.

If you know the size of the field of view of your eyepiece you can always make estimates of the extent of any object within that field of view. You should always compare the size of an object with the size of the field of view when you draw it, anyway (at least, that is the case when you draw the complete field of view within a representative large circle on your drawing paper).

The above estimate, can be made more accurate if you can more precisely determine the east–west span of any particular feature (the coma of a comet, just to take one example). You can do that by switching off the motor drive and letting the feature drift out of the field of view. You time from the moment the leading edge reaches the western edge of the field of view until the moment its following edge just disappears. The time taken for egress is t and the this value is used in $D = 15t \cos \delta$, the same equation used to determine eyepiece fields of view. In this case D is the east–west span of the feature you are wishing to measure.

What do you do if you want to measure the span of a particular feature in the north–south direction, or at some angle to the north–south or east–west directions? Here we are dealing with *position angles*. As Figure 4.6 illustrates, position angles are measured as compass direction radially away from a reference point. The position angle (PA) is defined to be 0° in a direction due north, increasing through 90° due east, further increasing through 180° due south, 270° through due west and finally to 0° (equivalent to 360°) once again.

How does one go about measuring position angles? You may be able to get an estimate accurate to within a few degrees from your drawing if you have the east–west or the north–south line accurately recorded on it. Letting a star track across the centre of the field of view and recording its entrance and exit points on your drawing will define the east–west line if you have an altazimuthly mounted telescope. If your telescope is mounted equatorially, and provided the polar axis is within a degree of its proper alignment, then clamping one axis and slewing through the other will do the same job. For instance, a star's apparent motion in declination, with the polar axis clamped, will adequately define the north–south line.

Another method that some people recommend is to carefully plot the star positions on your drawing and then use an atlas to identify the same stars, finally measuring the PA of the line between selected stars on the atlas. The line between the corresponding stars on

Figure 4.6 Position angles

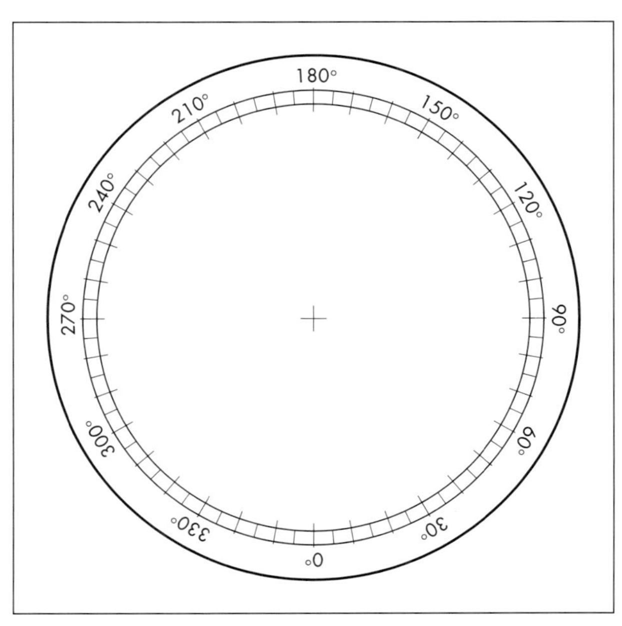

your drawing will have the same position angle and this can be used as a basis to measure other PAs on your drawing. In my view this method is clumsy and involves so many steps, with the potential for inaccuracy in each one of them, that the accuracy of the end result must surely suffer.

A more satisfactory procedure is to measure PAs directly on the field of view you see through the eyepiece. Once again the expensive, and nowadays rarely used, eyepiece micrometer is the best tool for this job. However, a number of manufacturers do produce eyepieces with a *reticle*, consisting of one or more engraved scales simultaneously in focus with the field of view of the telescope, which do allow measurements to be made.

At the time of writing the best of these is a new product marketed by the Meade Telescope Corporation. It is called a "MA 12 mm Illuminated Reticle Astrometric Eyepiece", and is advertised at a (2001) price of $120. Figure 4.7(a) illustrates the unit and (b) shows the

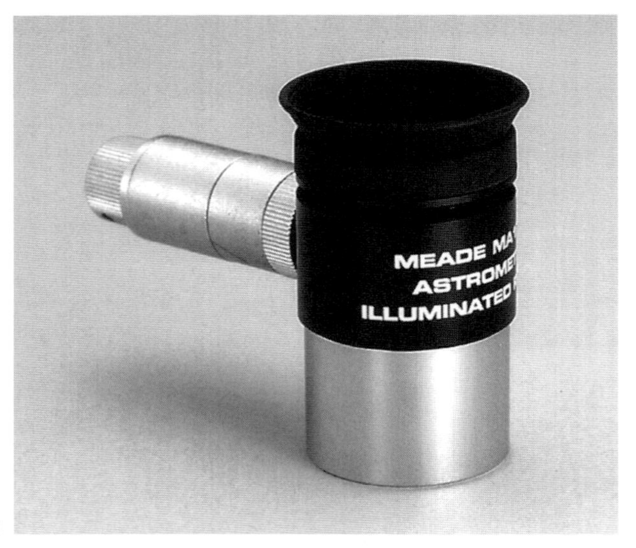

a

Figure 4.7 a The Meade Illuminated Reticle Astrometric Eyepiece. **b** The reticle the observer sees when looking through the Meade Astrometric Eyepiece. Note that the outer PA scale of this accessory to their range of catadioptric telescopes is designed for use with the addition of a star diagonal. In use with a normal astronomical telescope and no additional diagonal take the reading given by the scale but remember afterwards to convert it to a PA reading that corresponds to that shown in Figure 4.6. For example, a reading of 150° converts to an actual PA of 30°. Courtesy Meade Instruments Corp.
(Figure 4.7b. on the following page)

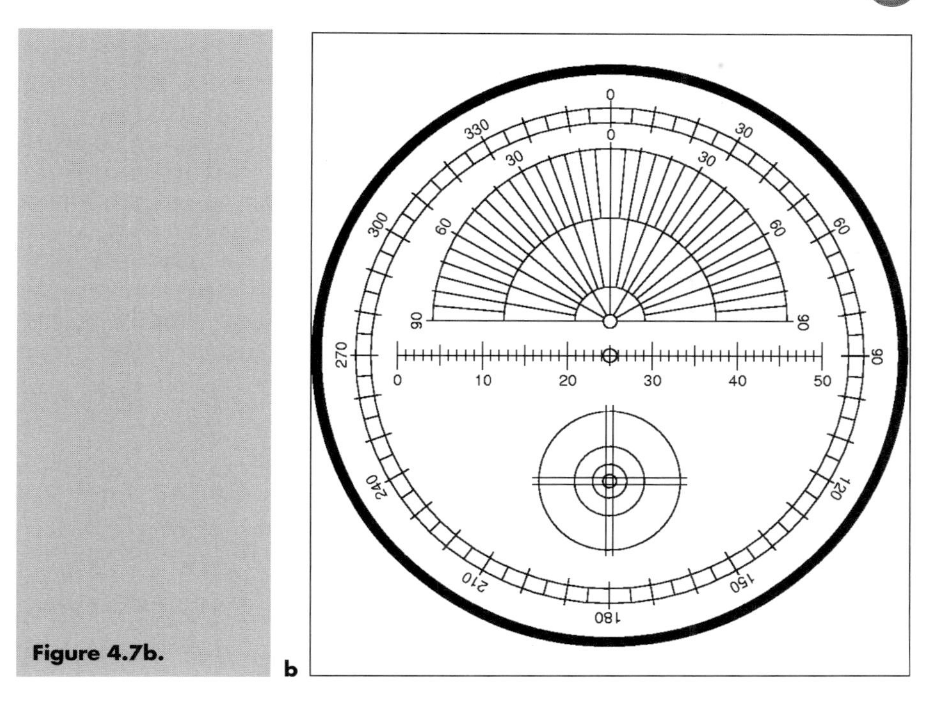

Figure 4.7b.

reticle that an observer sees superimposed on the field of view. The fact that the eyepiece type is the three-element "Modified Achromatic" will mean that it will not deliver an absolutely sharp image with telescopes of focal ratio lower than about f/6. Of course, that defect can be obviated by using a Barlow lens to increase the effective focal ratio of the telescope, though at the penalty of increasing the magnification with its attendant reduction in the size of the real field. However, absolute sharpness hardly matters when its function is really to measure lengths and directions of coarse details, not to critically examine the very finest details. To that extent it is an extremely useful accessory suitable for any telescope.

As you can see from Figure 4.7(b), the reticle contains more than just the outer PA scale. You could use the linear scale to measure lengths directly if you first calibrate the scale by the method of timing a star's drift (the telescope drive being switched off) along it. Having done that the unit can be rotated in the eyepiece drawtube until the linear scale then lies in the direction you wish to make the measurement. Reading the number of divisions covered by the feature you wish to measure and applying the calibration will allow you to determine the length of the feature.

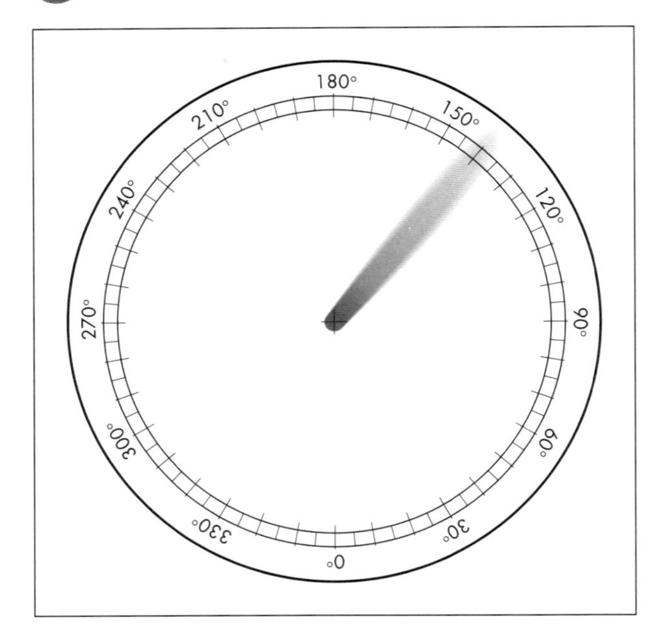

Figure 4.8 Measuring PAs using a simple reticle. In this example, the PA of the mid-line of the comet's tail is 140°.

To measure the PA simply use the outer scale in the manner illustrated in Figure 4.8, in this case with a much simpler reticle. To determine the length of a feature in a direction other than E–W using such a simple reticle, first measure the E–W length by timing the diurnal drift. Then read off the PA in order to determine the angle the feature makes with the E–W direction as shown in Figure 4.9. Then use trigonometry as shown in the same diagram.

Even a simple crosswire eyepiece will be of some aid, though the accuracy of your determined PA values will obviously not be as good. One thing that is mandatory for even the simplest reticle or crosswire eyepiece is some form of illumination. Otherwise you will not be able to see the scale, or crosswires, against the nearly black field of view. You might wish to have a go at making your own crosswire eyepiece but please choose a cheap, and preferably disused, eyepiece for your first attempt. I would hate to think of anybody ruining an expensive eyepiece because of anything I have written!

For measuring the PAs and lengths of features that are larger than the field of view of the telescope eyepiece, you could plot the two points (one at each end of the feature – for example the false nucleus of the comet and the end of the tail if you wish to determine the length and PA of the comet as a whole)

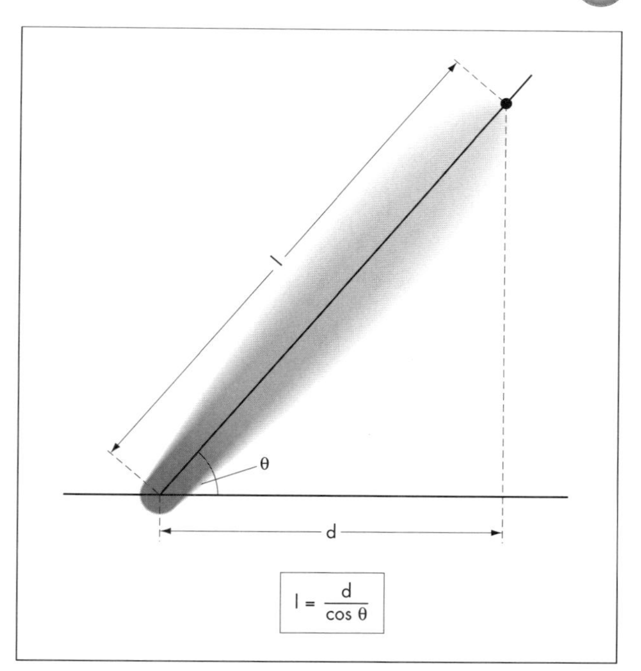

Figure 4.9 With the angle (θ) the feature (in this case the comet's tail) makes with the east–west line determined using a reticle eyepiece, and the span of the feature (d) in the east–west direction determined using drift, the formula can be used to determine the actual length (l) of the feature.

on a star atlas and then make your measurements on that. If the comet's tail is significantly curved then you should record PA values at the head of the comet and at the end of its tail.

There are other methods you could use. For instance, if your telescope has particularly accurate setting circles then these can be pressed into service (setting the telescope on each point and noting the readings). Choose whatever method will work for the equipment at your disposal and the particular comet at hand.

A final point: when submitting observations to the co-ordinator of an astronomical society, please do remember always to include **all** the details: your name, your address and the location of observation site (if different), the time(s), all details of the equipment you use – not forgetting eyepieces and magnifications used, notes to accompany all the results and drawings you submit, as well as any general notes of relevance, including brief descriptions of the methods you used to obtain measurements, etc. If your report lacks any of these details it is promptly diminished in value.

Chapter 5
Comets in Camera

On 27 September 1858 an English photographer called William Usherwood obtained the first ever photograph of a comet. He used an f/2.4 portrait lens and a seven second exposure on a collodion photographic plate. The resulting image of Comet 1858 VI Donati showed the bright nucleus and a faint tail and the photographer was sufficiently excited to send a copy to a leading astronomer of the time, Richard Carrington.

In 1858 photography was in its infancy and photographic plates were extremely slow with exposures of several seconds required to capture daylit scenes. Donati's comet was bright and it would have been an easy target for modern cameras and films. A few years ago many people secured excellent photographs of comets Hale–Bopp and Hyakutake using the simplest equipment. Sadly, such comets do not come along very often. Most that we can expect to encounter will be much fainter; difficult to see even with the aid of binoculars.

The following pages detail how you can undertake comet photography for yourself, starting simply and progressing to more effective, though more complex, techniques, and the use of electronic detectors. Comet photography through telescopes is, though, deferred until Chapter 7.

Most comets are faint, low-contrast, objects with diffuse tails or coma features which merge gradually into the sky background. Consequently, imaging a comet requires techniques similar to those used by deep sky photographers, though we have the added complication that comets move relative to the background stars. Since the techniques are so similar, you can use a suitable deep sky object to practice on if a comet isn't currently available.

Starting Simple – The Fixed Camera

If you have a sturdy tripod and a camera capable of taking time exposures you already have everything that is required apart, of course, from the comet! The very brightest comets can be photographed using fast film, a standard lens and a camera on a fixed tripod. This configuration is also suitable for fainter comets if they pass close enough to the celestial pole. Excellent photos have been taken using this arrangement (see Figure 5.1) but the opportunities are few and far between.

Even the brightest comets are rather faint by everyday standards and so it is necessary to use a very sensitive (or fast) film and an exposure of many seconds. Practically all Single Lens Reflex (SLR)

Figure 5.1. Comet Hyakutake in Perseus. 1996 April 14, $20^h 40^m$ UT. 30s exposure with a 50mm lens on a fixed tripod. T-Max 3200. COAA, Portugal. Stewart Moore.

cameras can give exposures this long if you use the "B" setting and they are ideal for astrophotography.

If you don't already own a suitable SLR camera, I recommend that you search for an old mechanical camera body on the second-hand market. You should be able to find something suitable for around £100/$150. For comet photography there is no need for features such as auto-focus or advanced automatic exposure systems. All that is required is a camera capable of taking manually controlled time exposures. Modern cameras generally have electrically activated shutters which drain expensive batteries during time exposures whereas older cameras rely on mechanical linkages. You will find that these low-tech cameras are far more reliable in the cold, damp conditions often found outside at night. The Olympus OM-1, long out of production, is often found on the second-hand market and it is an ideal camera body for astrophotography.

Although SLR cameras are ideal for our purposes, they may soon be a thing of the past. Film-based photography is being overtaken by digital cameras based on the magical CCD chip. Advanced forms of these cameras have revolutionised professional and amateur astronomy over the last decade. Unfortunately, while digital cameras use similar chips to the ones found in astronomical CCD cameras, most of the cameras you will see at your local store will not be suitable for comet imaging since they are incapable of taking long exposures.

Here I will assume that you have a 35mm SLR camera with a standard lens (which will have a focal length of about 50 mm) and that there is a bright comet hanging in the sky outside. To take your comet portrait you will need a locking cable release to hold the shutter open and a sturdy tripod that will not wobble during the exposure (Figure 5.2). Since comets are usually faint and relatively small you should use a fast film. A good choice would be one of the ISO 800 colour print films manufactured by Kodak or Fuji.

Place your camera on its tripod; set all the controls to manual; open the lens to its widest setting and set the focus to infinity. Point the camera at the comet; set the exposure setting to "B" and open the shutter for a time exposure of 10–30 seconds. That's all there is to it! However, to avoid disappointment, let me repeat that this simple arrangement will only work well on bright comets and you may have to wait several decades for a really spectacular subject to arrive.

Figure 5.2. All that is needed for simple comet photography is an SLR camera with cable release on a sturdy tripod.

The main disadvantage of this very simple technique arises from the unfortunate (in this context!) fact that the Earth rotates on its axis. This motion restricts the maximum length of an exposure on a fixed camera mount since too long an exposure will cause the stars to blur into trails. While star trails can make a nice "pretty-picture" they are not usually good for comet photography unless the comet is very bright and you have some interesting foreground detail (Figure 5.3).

The maximum exposure that you can use with a fixed camera is dependent on the focal length of your lens and the declination of the comet. All of this can be determined with some simple maths. At the same time you can calculate the field of view of your camera

Figure 5.3. Comet C/1995 O1 Hale–Bopp imaged using a fixed tripod on 1997 March 28. 2 minute exposure on Fuji Super G-800 film. Hale-Bopp was so bright that it could be photographed from the middle of towns. The 2 minute exposure leads to obvious star trails. Nick James.

system to ensure that the monster comet will fit in the picture. The image scale of any photographic system can be calculated using the simple formula given in Chapter 3. Here it is again with the constant changed so that the result is in degrees per mm:

$$Scale = \frac{57.3}{f}$$

In this equation f is the lens focal length in mm. From this you can see that a standard 50mm lens gives an image scale of just over 1 degree/mm. At this focal length the Full Moon is just about 0.5mm across and the field of view of a 35 mm film frame is roughly $41^\circ \times 27^\circ$. Very few comets in history have had an apparent tail length greater than this and even a bright comet may have a tail only a few degrees long. The vast majority of comets will appear very small at this focal length.

More important than the scale of the image on the film its scale on the final print. If you enlarge your 35 mm negative to an 8 inch × 10 inch (200 mm × 250 mm) print (an enlargement of about 8) your 50 mm lens will produce an image scale of approximately 0.125 degree/mm. As discussed in Chapter 4, the Earth's rotation introduces an apparent movement of the sky corresponding to $15 \cos \delta$ arcsec/sec where δ is the comet's declination. An object on the equator ($\delta = 0°$) will move 0°.125 (450 arcsec) in 30 seconds and so stars will appear as 1 mm streaks on the print. This will be quite noticeable and so you should keep fixed exposures with a 50 mm lens to less than 30 seconds. Longer focal length lenses demand proportionately shorter maximum exposure times.

You can see that a fixed tripod will limit your exposures and so it is important to use a fast film to record the maximum detail in the comet's coma and tail. Luckily film technology has given us some excellent fast films with relatively fine grain. Colour print films such as Fuji Superia X-TRA 800, Kodak Supra 800 and Kodak Zoom 800 will prove ideal for this type of photography. Bright comets are often at their best when seen in a twilight sky and so you can often add interest to your images by including foreground detail. Experiment with exposures and don't skimp on film. Bright comets don't come along very often and a good picture is worth the cost of a few rolls of film. As with your visual observations it is important to record the date and time and other important details such as the film used, the focal length, image scale and orientation.

If you are new to astrophotography you will probably be using the standard lens on your camera. Since comets are faint and your exposures are limited it is necessary to use the lens wide open (the smallest "f-number" setting). However, this will show up various optical shortcomings in all but the very best lenses and you may well find that the stars near to the edge of your field of view look more like seagulls than dots. You could experiment with the lens set to one stop below its maximum aperture to see whether it improves the optical performance but you will probably find that the loss of faint detail will outweigh the benefit of sharper star images.

Sometimes longer exposures are possible with a fixed camera. On occasions nature conspires in our favour and a comet moves close to the pole of the sky. When this happens $\cos \delta$ becomes small and it is possible to expose for longer since the apparent motion is less. This occurred for Comet Hyakutake in 1996 as it passed through Ursa Minor (Figure 5.4) but the most noteworthy case was Comet LINEAR (C/1998 M5) in 1998 which passed so close to the pole that a 50mm focal length fixed-camera exposure of three hours was possible. For several hours on 1999 March 15 the motion of this comet relative to the background stars almost cancelled out the small apparent rotation of the sky, so the comet was effectively stationary. Between 10^h and 11^h UT the comet actually moved less than two arcseconds in the view of a fixed camera! This was one of those very rare occasions when a fixed camera could image a rather faint comet and medium focal lengths could be used without a drive. If you are restricted to using a fixed camera it is always worth watching out for these opportunities.

So, the fixed camera arrangement is effective in certain circumstances. It is though, rather limiting and you will find it useful on only a few comets each decade. Fixing your camera to a mount which can move to

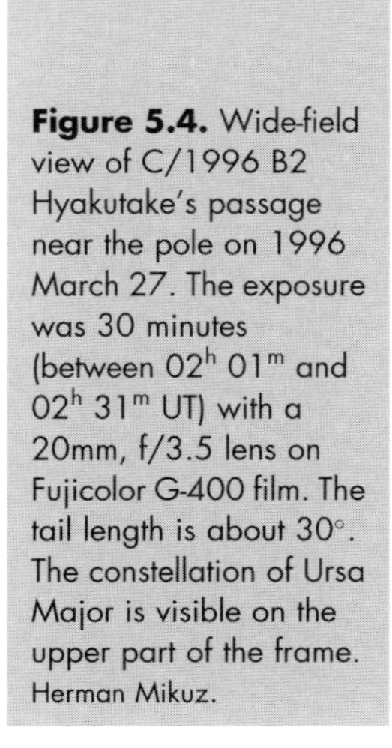

Figure 5.4. Wide-field view of C/1996 B2 Hyakutake's passage near the pole on 1996 March 27. The exposure was 30 minutes (between $02^h\ 01^m$ and $02^h\ 31^m$ UT) with a 20mm, f/3.5 lens on Fujicolor G-400 film. The tail length is about 30°. The constellation of Ursa Major is visible on the upper part of the frame. Herman Mikuz.

counteract the Earth's rotation will free you from this restriction. Such a mechanism is easy to build and it will allow you to photograph much fainter comets wherever they are on the sky.

We will return to the practical details of comet imaging shortly but it is now worth taking a short detour to look at the different kinds of image detectors available.

Film or Chips?

Any camera consists of two elements: an optical assembly of some kind which forms an image on a light-sensitive detector. Until the 1990s most imaging used light-sensitive film as the detector and while there have been tremendous advances in film technology over the years it is still based on chemistry which is similar to that used by Underwood for his first comet photo. Silver halide (chloride or bromide) is layered into a gelatine medium on a transparent base. At one time glass provided the base, but later celluloid, then acetate plastic was used. So, the old rigid "plates" gave rise to the flexible films that are so familiar today. Light falling on the film liberates the silver ions from their bonds with the halide ions and these are converted into grains of silver when the film is developed. The brightest parts of the image give rise to the darkest parts on the developed film. So, the film is a *negative*. Positive prints can then be made by shining light through the negative onto light sensitive photographic paper (the darkest parts of the negative then giving rise to the lightest parts of the final print). This reversal of tones can be undertaken photochemically on the film itself to produce a *reversal* or transparency which can then be projected onto a screen.

Competition between manufacturers ensures that film brands are continually changing and at any one time there are always a few films which are particularly suitable for comet imaging. Table 5.1 presents very limited list of the types of films available at the time of writing but bear in mind that the list is likely to be out of date before you read it!

In the early 1970s developments in semiconductor technology gave rise to a number of electronic marvels. The first microprocessors appeared and these were developed into the powerful computers that we use

Table 5.1. Suggested films for 35mm photography. Bear in mind that colour films come and go rapidly and that this information may be out of date by the time you read it! Almost any modern, fast colour film will perform very well.

Kodak Tech-Pan	B&W	An excellent,lent, high-resolution film which requires hypersensitization for comet work.
Fuji Super G800	Colour print	Very sensitive, good colour balance and relatively fine gain. Gave excellent results on Hale–Bopp and Hyakutake. The more recent Fuji Superia 400 also gives excellent results.
Kodak Elitechrome 200	Colour slide	Low reciprocity failure and very fine grain. Excellent for bright comets with fast optics and reasonably long exposures. Responds well to push processing (i.e. increasing sensitivity at the development stage).

today. At about the same time the Charge Coupled Device (CCD) made an appearance. CCDs are essentially large, two-dimensional arrays of light-sensitive elements (Figure 5.5). Each element is called a pixel and an array of pixels records the image. The resolution and imaging area of a CCD is defined by the number of pixels it has and the size of each pixel. The smallest CCD imagers have 256 pixels or less on each side and measure only a few mm square. At the other end of the scale there are now CCDs available which are bigger than a 35mm film frame and which contain in excess of 16 million pixels.

A CCD performs the same job as a piece of film except that the image is collected and stored electro-

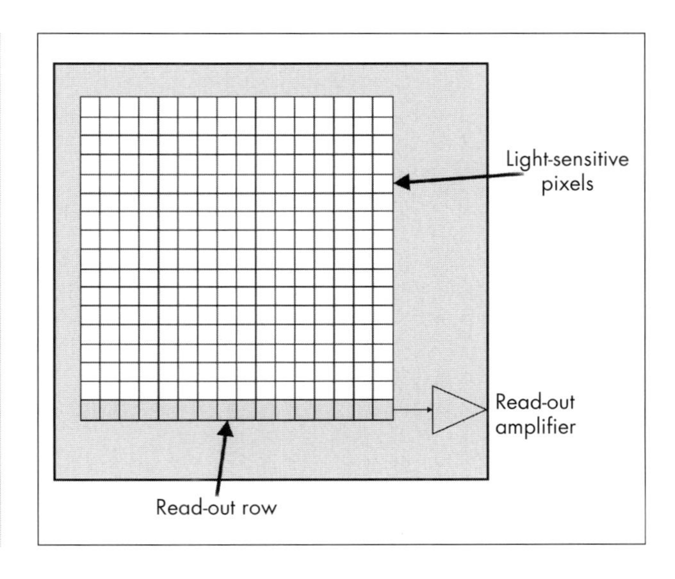

Figure 5.5. This diagram shows how the light-sensitive pixels are arranged in a full frame CCD. At the end of the exposure the charge is shunted down the columns and then read out, row-by-row, to form the image.

nically rather than photochemically. It is expensive to manufacture large CCDs and so most commercially available devices are considerably smaller than 35mm film although they may still have large numbers of small pixels. The pixel structure of a typical CCD is quite complex but for our purposes it can be considered to be a storage well for electronic charge. Incoming photons interact with atoms in the well and cause charge to be generated in proportion to the intensity of light falling on the pixel. This charge accumulates in the well as the exposure continues. Unfortunately there is another source of charge which trickles into the well during the exposure. This thermally generated charge is a form of noise, or unwanted signal, that can swamp the desired signal but it can be reduced by cooling the chip.

A CCD integrates charge in the pixel wells in a similar fashion to the way that film builds up silver ions in proportion to the light falling on each grain. Ignoring the thermal noise, the charge in each pixel well is linearly proportional to the product of the exposure time and the light intensity falling on that pixel.

Like a real well, the pixel well in a CCD has a limited capacity and bright objects can cause it to overflow. A feature of CCD chips which has a great relevance to us is the *anti-blooming drain*. Without this feature bright objects such as stars can cause the charge in the affected pixel to overflow into adjacent pixels. The effect is that the charge overflows down the columns causing a vertical line centred on the star (Figure 5.6). If you are trying to image a faint comet which has bright stars appearing as if they were embedded in the tail this effect could ruin your images. Many modern CCD chips and practically all of those used in digital cameras have "anti-blooming" drains built into the pixel structure. These drains act in the same way as an overflow on a bath tub and the excess charge flows harmlessly away into the semiconductor substrate. As you might expect the benefits of anti-blooming drains do not come without a price and they can reduce the linearity of the pixel's light response (that is twice the image brightness does not correspond to twice the accumulated charge at the end of an integration). This is why they are often avoided in professional astronomical CCDs and you should avoid them too if you plan to do photometry with your camera as described in Chapter 9.

An ideal CCD would turn each incoming photon into a measurable unit of charge. Sadly, this is not the case for real devices. The sensitivity of a CCD, or the

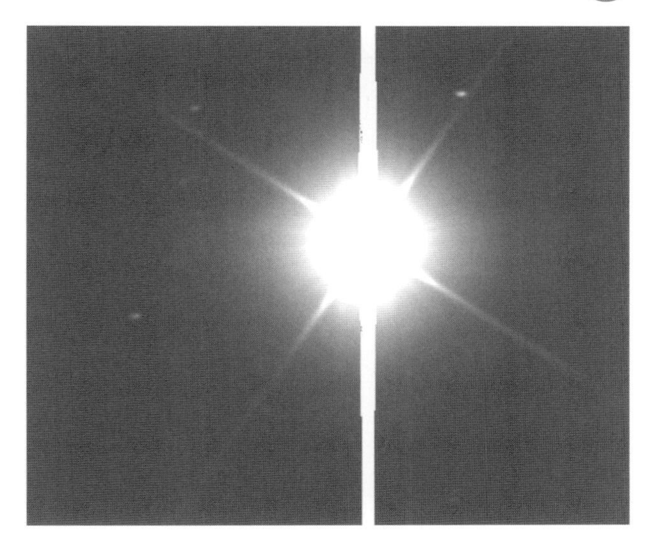

Figure 5.6. This image of α Orionis (Betelguese) was taken using a CCD camera which does not have anti-blooming (AB) drains. The charge from the star has overflowed and filled an entire column of pixels (the other spikes from the star are diffraction spikes caused by the optics). Digital cameras usually include AB drains to avoid this effect but they reduce the sensitivity

efficiency with which it collects each photon falling on its front surface, is called the *Quantum Efficiency* (or QE). A number of factors conspire to reduce the QE of a CCD chip. It may be that the photon does not have enough energy to release a packet of charge. Lower energy photons correspond to longer wavelengths and so this effect causes a drop off in sensitivity towards the red end of the spectrum. This loss of sensitivity actually occurs well beyond the deepest red that can be seen by the human eye and most CCDs are sensitive well into the infra-red. At the other end of the spectrum high energy photons may not get deep enough into the well to liberate any charge and so sensitivity also falls off towards the blue. Some QE curves for common CCDs are shown in Figure 5.7. You can see that some CCD chips have very high QE values in excess of 90% over a wide range of wavelengths but you can also see that different CCDs have radically different responses. Even the very best films have a QE of around 5% and this shows the huge advantage that CCDs have over that older medium.

There are three common CCD structures found in modern cameras. Full-frame CCDs are the simplest and they are used for professional astronomical imaging. During the exposure charge is collected in the pixel wells and at the end of the exposure individual packets of charge are shunted down the vertical columns into the horizontal register. The charge packets in the horizontal register are then shunted out one by one via

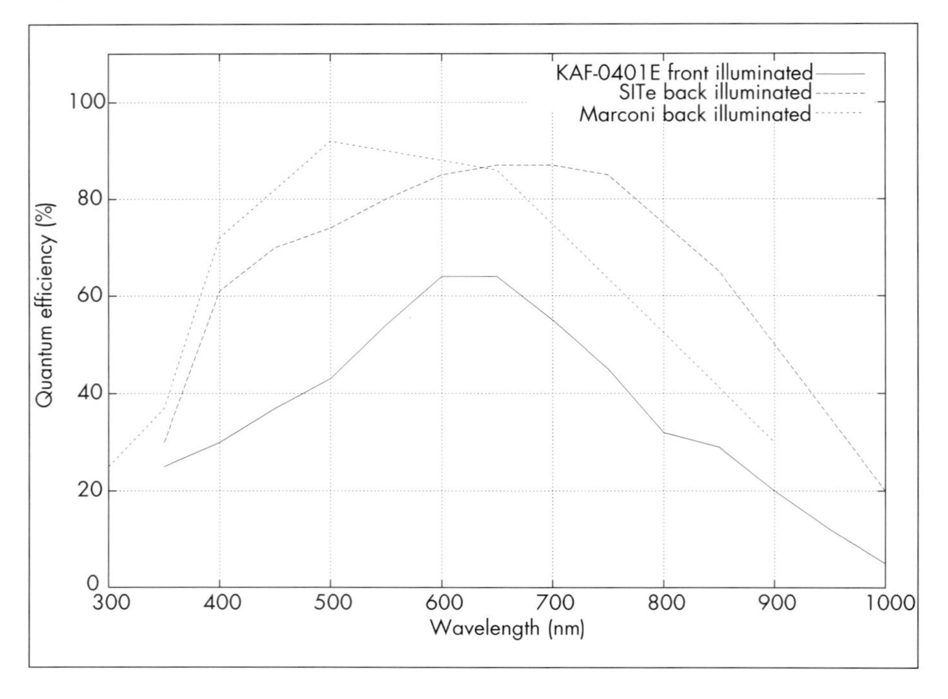

Figure 5.7. This graph shows how the response of various common CCD chips varies with wavelength. The best devices have Quantum Efficiencies in excess of 90%. The very best film detector only achieves 5%.

the sense amplifier where the amount of charge is measured and represented as a digital number which can be stored in a computer. In a 4K by 4K chip there are over 16 million charge packets and each one must be shunted to the sense amplifier, converted to a digital number and stored. The amazing thing about the CCD design is that charge can be shuffled down the columns and then along the rows with very little loss. As each packet reaches the sense amplifier it is converted into a digital number for storage and subsequent processing. The resolution of this number is usually 8 bits (2^8 or 256 shades of grey), 12 bits (4096 shades) or 16 bits (65536 shades). When dealing with low contrast subjects such as comets your camera should have at least 12 bits and preferably 16 bits of resolution. The read-out time of a CCD is an important parameter and this can be many seconds or even sometimes minutes for large CCDs. This dead time between exposures can be a problem if you are taking multiple consecutive images of a comet which will subsequently be combined into a single image.

One disadvantage of the full-frame CCD is that the light sensitive pixels are used to store charge while the image is being read out. If the CCD is exposed to light during read-out the image is smeared as the charge is moved down each column. To overcome this most astronomical CCD cameras include a mechanical shutter to ensure that the chip is not exposed to light during readout. In an alternative structure, called a frame-transfer CCD, half the pixels are covered with an opaque screen. At the end of the exposure the image is quickly transferred from the light sensitive area to the masked area. It can then be read out in a leisurely way with no need to worry about image smear and no need for an expensive mechanical shutter. The disadvantage with this technique is that only half the area of the chip is available for imaging.

A radically different structure is popular in low-cost commercial CCDs designed for video cameras. In the *interline* structure each column of light sensitive pixels has an associated column of storage registers which are covered with an opaque mask. The pixels are exposed in the normal way but at the end of the exposure the packet of charge in each pixel is dumped into the adjacent interline register. The image can then be shifted out in a similar way to that used in full-frame cameras but with the added advantage that the light sensitive pixels can be collecting a new image while the old one is being shifted out and stored. This is rather useful when taking multiple, successive, images since it eliminates the "dead-time" that is associated with read-out in full-frame CCDs. Interline CCDs are generally less sensitive than the full-frame or frame transfer varieties since the light-sensitive part of each pixel is a small proportion of the total pixel area. Various manufacturing techniques have been used to improve this sensitivity but interline CCDs are not often found in high quality astronomical CCD cameras.

CCDs vs. Photographic Films

Both CCDs and conventional film are analogue image detectors. The difference between the two is that the image data is stored in situ as an analogue signal in film but it is immediately converted to a digital signal during CCD readout. After chemical development a scanner can be used to convert film's stored analogue image into a digital representation. Consequently film and CCD images can, if desired, both end up in the same format and can be processed in the same way. The advantage of a CCD is that the image is available far more quickly without all the intermediate chemistry.

Film has the advantage that a large imaging area can be obtained at low cost and there is no need for supporting batteries or electronics which can fail in the cold. A fine-grain 35mm film has a resolution of around 200 pixels/mm and so the 36 mm × 24 mm frame is equivalent to a huge CCD with 7200 × 4800 pixels. CCDs are usually much smaller than 35mm film so the enlargement to the final print scale is much greater. An 8 inch × 10 inch print is only an eight times enlargement of the original 35mm negative but such a print might be 20 or 30 times the size of the original CCD. The key requirement for a "photographic quality" final print is that you need a resolution of 150 dots per inch (dpi) or better at the final print scale. A CCD detector would require in excess of 1500 × 1200 pixels to achieve this for an 8 inch × 10 inch print. The actual physical size of the detector does not matter except that it defines the focal length and quality of the optical system to give a certain field of view and resolution.

A further advantage that film has over digital or CCD cameras is initial cost. It is true that the running costs of film-based photography are higher but the initial investment is much lower. Compared to most of the CCD cameras that are suitable for astronomy, the common SLR is highly portable and does not require a power supply or external computer.

Film does have a number of disadvantages compared to CCDs. Even the best films are not particularly sensitive and since film can only be used once the running costs of film cameras can be much higher their electronic replacements. While there have been great improvements in the last few years the *dynamic range* (or the difference between the brightest and darkest areas that can be correctly recorded for a given exposure) is limited compared to that available in CCDs. This is particularly important for comets where details in the inner coma can be drowned out by over exposure if you expose to capture the much fainter tail (Figure 5.8).

Film also suffers from an effect called *reciprocity failure*. An ideal detector would have a reciprocal relationship between the required exposure length and the brightness of a scene. In normal photography, when exposures are short, this is indeed what happens and you would use twice the exposure length to record a

Figure 5.8. This CCD image of 122P/de Vico taken on 1995 October 5 has been processed to demonstrate the limitations of photography. The contrast has been stretched to bring out the faint tail structure but this means that the coma is saturated. Any interesting detail in that region is lost. CCDs have an advantage over film that bright and faint regions can be recorded simultaneously and then displayed by a non-linear contrast stretch. Denis Buczynski.

scene which is half as bright. As the scene brightness falls there comes a point when the exposure must be extended by more than would be predicted by this simple rule. This reciprocity failure means that the effective speed of the film when used with long exposures is considerably less than the ISO rating printed on the box. In most astronomical situations the light level is so low that reciprocity failure is a serious problem.

Dedicated comet photographers can reduce the effects of reciprocity failure in some films by using a technique called *hypersensitisation*. This involves baking the chosen film for many hours under pressure in Forming Gas (a mixture of Nitrogen and Hydrogen). Some intrepid photographers have even used pure Hydrogen but this is only recommended if you really know what you are doing! The effects of hypersensitisation can be spectacular especially when used with films such as Kodak's Technical Pan black and white film. It is even possible to buy film which has been hypersensitised (see the Appendix) although it is becoming harder to find as CCDs gradually replace film for astro-imaging.

CCD detectors do not suffer from reciprocity failure but they have their own problems when used for long exposures since thermal noise can be troublesome. The thermal noise builds up during each exposure and it can only be removed using a *dark frame* if the exposure is kept short enough so that there are no saturated pixels in the image. A dark frame is an exposure which is the same length as the one used for the actual image but which is made with the shutter closed or the lens capped. The only pixel charge in a dark frame comes from the thermal noise and so it can be subtracted from the raw image to remove the thermal noise effects. Even with dark frames it is necessary to cool the CCD detector (often to −30°C or below) but this is expensive and is unnecessary in digital cameras intended for normal photography.

Probably the biggest advantage of CCDs compared to film is their ability to cope with considerable local light pollution. The dynamic range and linearity of a CCD means that it is possible to subtract the sky background light from the final image to reveal objects which are perhaps only 1–2% as bright as the sky. I will have much more to say about these aspects in Chapter 8.

The chemical processes involved with film work can be messy and time-consuming. Darkroom technique could fill an entire book on its own and I will not discuss the subject here. In any case the techniques of digital photography are now being applied to film and many astrophotographers never go anywhere near a darkroom. It is quite easy to get your negatives developed and scanned commercially with the resulting files returned on a CD-ROM. You can then process the images on a computer in the same way as those obtained using CCDs. I certainly find that digital image processing on a PC is much more pleasant than the work I used to have to do in a darkroom. You can try things out quickly and reject any mistakes without breaking the bank balance and some techniques which were extremely difficult and time-consuming in the darkroom are very straightforward when performed digitally. Digital image processing is a powerful tool and it will be covered in a lot more detail in Chapter 8.

Despite continual warnings of the death of film-based photography there is still a great deal of it about and films are better now than they have ever been. The combined attributes of high speed, high resolution and low reciprocity failure have been achieved in certain colour emulsions. On the other hand digital cameras and CCDs will no doubt improve and fall in price so that, one day, they will completely replace the humble 35 mm SLR. Whichever detector you decide to use for comet imaging you can be confident that the techniques described here will be applicable.

Which Lens Should I Use?

In addition to the light-sensitive element any camera system must also include an optical arrangement to bring an image into focus on the detector. Since comets are extended objects their apparent brightness on the film or CCD detector will depend on the focal ratio of the optical system (assuming that the focal length of the system is large enough that the comet is shown as a fuzzy, extended object on the image). The faster the lens (i.e. the smaller the f-ratio) the better. Most comets are small and so quite long focal length lenses are required for good results. The combination of long focal length and fast f-ratio can be an expensive one. Still, good results are possible with a humble 135 mm, f/2.8 lens.

As I mentioned earlier, the standard 50mm lens gives a field of view of roughly $27^\circ \times 41^\circ$ on 35 mm film. The field of view will of course be narrower if a smaller CCD detector is used. It is only very rarely that comets exceed these dimensions. Hyakutake (C/1996 B2) had a tail length in excess of 50° at its closest approach but most other relatively bright comets have tails only a few degrees long. In these cases focal lengths of 135 mm to 300 mm are a better choice if you are using 35 mm film. In all cases remember that the density of the image will be dependent on the focal ratio and you should use the lens at its widest setting consistent with reasonable optical quality.

Most lenses will have a relatively poor optical performance when used at their widest aperture. If you are using a sidereal drive it may be worth closing the iris by one stop to improve the optical performance and extending the exposure to compensate. An alternative is to pay a high price for a lens which performs well when used wide open. The picture of Hyakutake shown in Figure 5.9 was obtained using a Canon 85 mm, f/1.2L aspheric lens. This type of lens has a very good performance when used wide open, even at the edge of the field, and it is still available on the second-hand market for around £650. Similar lenses are available from various manufacturers but they are seldom justified by comet photography alone.

To get sharp star images the lens should be focused to infinity but you may find that the infinity mark on the lens isn't actually the best focus point and some experimentation will be required. This is particularly true with long focal length mirror lenses which often have no stop at the infinity point. Since different wavelengths of light are brought to focus at different points the infinity focus mark usually only applies to light in the visible range. Many CCDs are sensitive well into the infra-red and it is often necessary to employ an infra-red blocking filter to get sharp star images.

An ideal lens would provide even illumination over the entire focal plane so that the sky background would produce a uniform density on the film or CCD detector. Most lenses do not achieve this and the amount of light reaching the edge of the detector is less than the amount received by the centre. This *vignetting* effect was mentioned in Chapter 3 and if it is not corrected it will lead to an uneven background in the final print. This is particularly troublesome in the case of a large, low contrast object such as a comet's tail since detail will be lost when you try to print the final image with a high contrast.

Figure 5.9. Comet C/1996 B2 Hyakutake from Mount Teide, Tenerife. 1996 March 25, $00^h\ 19^m$ UT, 4 min exposure using a Canon 85 mm, f/1.2L lens stopped to f/1.8 and hypered 35 mm Kodak Tech Pan. The image shows a great deal of detail in the tail including a number of disconnection events. The field of view is $24° \times 10°$ and η UMa is the bright start near the top centre. Nick James, Martin Mobberley and Glyn Marsh.

Luckily, the effects of vignetting can be removed using flat fields. These are exposures of a uniformly illuminated subject (such as the sky) which are subsequently used to correct for various effects including the vignetting of the optical system. I will discuss these in more detail in Chapter 8. It is even possible to generate a synthetic flat field from a knowledge of the lens performance. While vignetting correction can be performed using complicated dark room techniques it is much easier to do on a PC when the images are in digital form.

Some very impressive comet images have been taken using medium format cameras and lenses of 200–400mm focal length. Such combinations are very expensive but if you already have the required equipment it will be ideal for driven and guided cometary photography. Medium format cameras have a combination of generous image scale and wide field which is very advantageous when imaging fine detail in a large comet.

Driven Camera Platforms

You have seen how a fixed camera will limit your exposures to a few seconds. Things are much better if you can mount your camera on a drive of some sort. Most astronomical driven mounts are based on the *equatorial* arrangement. The main (or polar) axis of the drive is arranged so that it aligned with the Earth's axis. The camera is then driven at the sidereal rate around this axis to follow the apparent motion of the stars. Some commercial computer-controlled telescopes use an altazimuth drive system but such drives are not suitable for wide-field imaging. The altazimuth drive follows the stars using complex motions in both axes and this leads to an apparent field rotation during the exposure. While mechanical *de-rotators* are available for prime-focus imaging they are not suitable for wide-field imaging using a camera.

Once you have a system which allows you to counteract the apparent rotation of the sky you will be able to take exposures of many minutes or even hours. Such images, obtained with short or medium format lenses, can be really spectacular, showing many stars, nebulae – and comets! Unlike the case for fixed photography, the exposure time for driven photography is limited by inaccuracies in the drive or by sky-fogging caused by man-made light pollution or natural twilight. Sky fogging is dependent on the speed of film used, the lens f-ratio and the sky brightness. If you live in a town you might be limited to exposures of only a minute or two at f/2.8. At darker sites in the country exposures of half an hour or more are possible. As always, experimentation is required.

Since comets are best photographed from dark locations you will probably have to travel out into the countryside to get good images. If this is so then your camera mount should be portable. Commercial mounts such as the Vixen SP and GP are of excellent quality and I have used one of these to image Comet Hyakutake from Tenerife. They can carry heavy cameras and ancillary equipment such as guidescopes but they are relatively expensive (around $1K for a motorised version). Alternatively it is quite possible to build your own drive. A driven mount for wide field imaging does not need to be very complex and the Scotch (or barn-door) camera mount is ideal. This can be built at home with very little skill and nothing more complicated than two planks of wood, a hinge and a screw (Figure 5.10). The screw rotates at just the right speed to force the upper plate around the hinge axis at the sidereal rate. You will find a description of a more advanced version of this kind of mount in the book *Advanced Amateur Astronomy* and I have used this to obtain spectacular photos of the recent great comets. Such a mount is compact and easy to carry to your favourite observing site.

In 1996 I took my simple Scotch mount to Tenerife to get a clear view of Comet Hyakutake. Figure 5.9 shows an example of the pictures I obtained using nothing more than an SLR camera and a simple barn-door mount. Hyakutake was an excellent example of a comet with a great deal of fine detail in the tail. You can see how easily that detail would have been lost if the drive were not accurate. The mount allowed me to use exposures of 4 minutes with an 85mm focal length lens and you should be able to achieve the same performance if you build the mount carefully. More complex driven mounts will allow exposures of hours.

The simple techniques involved with short focal length imaging of comets will produce impressive results on bright comets but such objects are quite

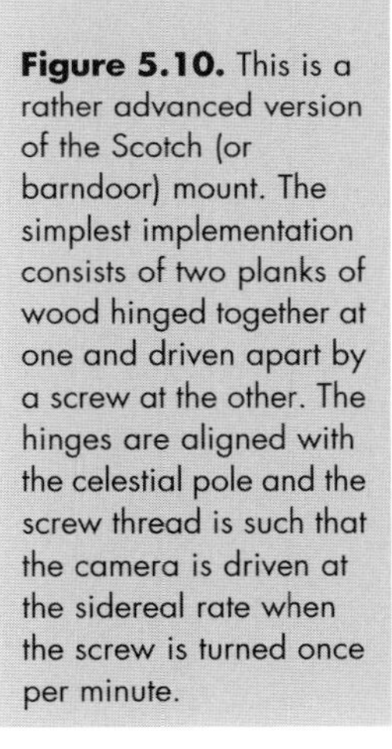

Figure 5.10. This is a rather advanced version of the Scotch (or barndoor) mount. The simplest implementation consists of two planks of wood hinged together at one and driven apart by a screw at the other. The hinges are aligned with the celestial pole and the screw thread is such that the camera is driven at the sidereal rate when the screw is turned once per minute.

rare. Much more common are the faint, fuzzy comets that lurk just below naked-eye visibility. There are usually a few of these in each year and so they provide us with a much more regular target for our cameras. This type of comet might have a tail up to a degree or so in length and so they would be best photographed using longer focal length instruments (Figure 5.11). Focal lengths of 500 mm or so (for 35mm film) or 135 mm for most CCDs would be about right here. At this focal length the simple barn-door mount is unlikely to track accurately enough and even some commercial drives might not be up to the task. Furthermore, at these focal lengths the motion of even quite faint comets can become appreciable and so some form of corrective guiding is required.

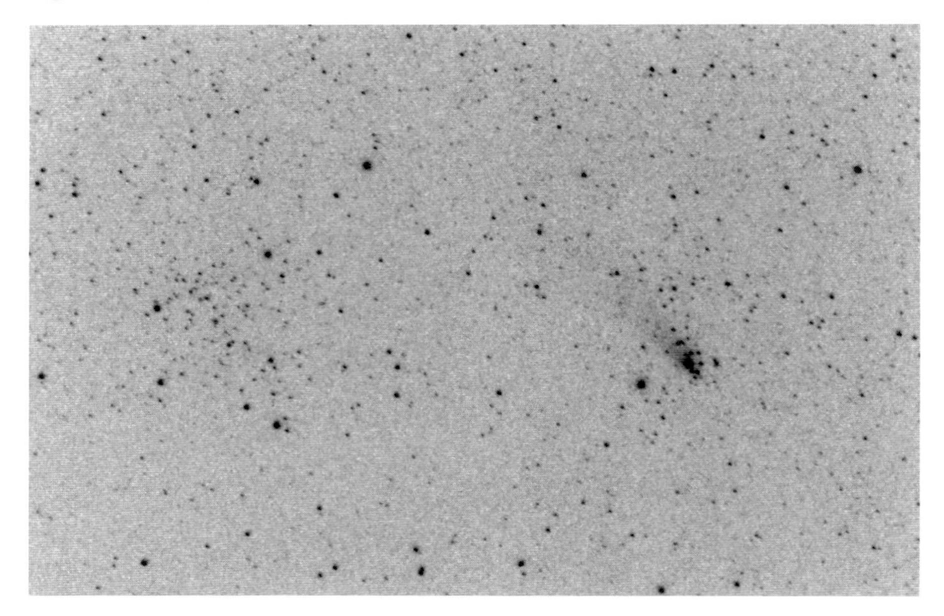

Figure 5.11. Comet C/1987 P1 Bradfield crossing NGC6633 on 1987 November 14. This is a detail from a 35 mm negative obtained using a 2 minute exposure, 135 mm FL lens at f/2.5 on Kodak T-Max 400. The camera was mounted on a Scotch mount shown in Figure 5.10. At the time this comet was at magnitude 5. This type of comet is near to the limit of this kind of simple equipment.
Nick James.

Guided Imaging with Medium Focal Lengths

Our next upward step in complexity is to attempt longer exposures with longer focal lengths. At focal lengths of 500 mm or more even expensive drive systems may not be accurate enough to counteract the Earth's rotation for more than a few minutes. In addition, at these focal lengths, the comet's motion against the star background must be considered. Indeed, this motion may be a factor even at short focal lengths if the comet is nearby. In 1983 Comet IRAS–Araki–Alcock was travelling so quickly that it could be seen to move against the star background when observed at a moderate magnification. At its closest a driven exposure of a few minutes with a 135 mm lens would have showed the comet as a streak against the fixed background stars (Figure 5.12).

If you are using an electronic camera you can overcome the problems of drive inaccuracy and comet motion by taking multiple short exposures. These can then be digitally combined after aligning the comet in each image. Life is much more complicated if you are using film since it will be necessary to guide the camera

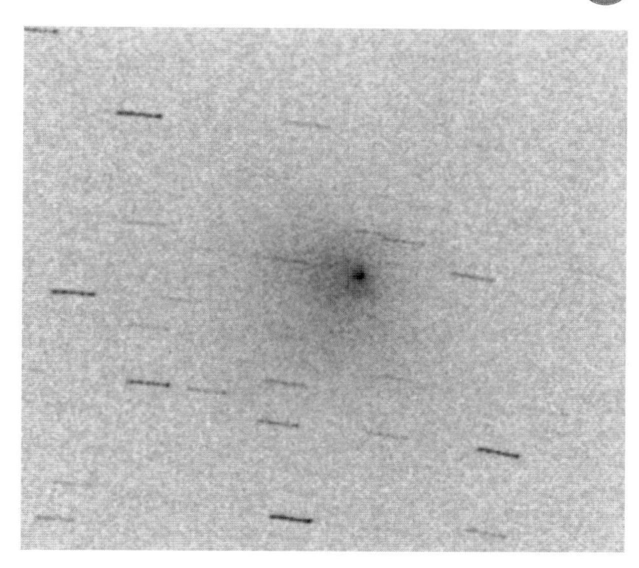

Figure 5.12. Comet C/1983 H1 IRAS–Araki–Alcock was large, diffuse and fast-moving. It came closer to the Earth than any other comet since Lexell's Comet in 1770. This picture was obtained on 1983 May 9 between $21^h\ 11^m$ UT using a 300mm, f/5.6 lens and Kodacolor 400 Film. Martin Mobberley

to follow the comet's motion. A common technique is to mount the camera parallel with a telescope on an equatorial mounting so that the camera does not look through the telescope but the observer does! This is something of a half-way house between the simple unguided, short focal length imaging that we discussed earlier in this chapter and the long focal length, telescopic imaging that we will discuss in Chapter 7. An equatorially mounted telescope with slow motions in both axes is required and remember that a computer-controlled altazimuth mount will not do since there will be a problem with field rotation.

If you are simply correcting for drive inaccuracy, all that you need to do is find a relatively bright star that you can see through the telescope while the camera is pointing at the comet. Use a high powered eyepiece with an illuminated reticle (such as the on described in Chapter 3) and place the guide star at a known point. You must then use the slow motions to keep the star on the crosswire for the duration of the exposure. If you have a good drive this can be a fairly relaxing activity but if your drive is unreliable you will need a great deal of concentration. An electronic timer can be set to "beep" to indicate the end of the exposure. The magnification that you use should be the highest that you can get away with while still having a clear view of the star. Of course there should be no flexure between the camera and guide telescope but basic guiding is relatively straightforward. If, in addition to correcting the drive, you are attempting to compensate for the comet's motion things become a lot more complicated and I will defer discussion of *offset guiding* until Chapter 7.

Can I Use My Digital Camera?

As mentioned earlier, the image from a digital camera is stored in a file which can be processed on a computer. Specialised astronomical CCD cameras usually require an external computer and other support equipment and I will discuss these in Chapter 7. Astronomical cameras are quite expensive but they are designed with the specific purpose of imaging very faint objects with long exposures. The detectors are cooled and the readout is slow with an emphasis on high accuracy. A quick glance at the glossy astronomical magazines will show that there is a range of cameras available covering a very wide price and performance range. If you have such a camera it is possible to attach a short focal length lens to take wide-field images of the night sky using long exposures. Some spectacular images of comets have been taken in this way (Figure 5.13).

Mass-produced digital cameras use the economies of scale to reduce the price of a very sophisticated unit to much lower level than their specialised astronomical cousins. In fact astronomical CCD cameras and mass-produced digital cameras often use similar CCD detectors but the digital cameras are considerably cheaper. Digital cameras are also self-contained since they have built in display and image storage systems. On the downside they lack detector cooling and most cameras in the amateur range do not have interchangeable lenses nor the ability to take manually controlled

Figure 5.13. Wide-field mosaic image of comet Hyakutake, taken on 1996 April 14 with a 180 mm f/2.8 lens, CCD and narrow-band H_2O^+ filter, centered at 620 nm. Two consecutive frames were taken between $19^h\ 03^m$ and $19^h\ 20^m$ UT. Each frame was exposed for 5 minutes. The frame field of view is $6^\circ.6 \times 2^\circ.4$. Herman Mikuz.

time exposures. They also tend to store images using data compression and even in "high quality" mode there is an inevitable loss of information.

With all of these disadvantages is it still possible to use digital cameras for comet photography? The answer is yes but only if you have one of the very few which allow time exposures of many seconds. Demand for an SLR replacement at the amateur level means that some digital cameras suitable for night-time imaging have come on to the market. The performance of these cameras is surprisingly good even when used at very low light levels. Probably one of the most impressive digital SLRs is the superb Canon EOS-D30. This is capable of taking long exposures using standard interchangable Canon EF lenses. The image sensor is huge (2160 × 1140 pixels) and the camera costs around $2K. Lower down the price range is the Nikon Coolpix 990. This also allows long exposures and a number of astrophotographers have used it with success.

The simplest arrangement is to place your digital camera on a fixed tripod. As with a film camera your exposures will be limited to a few seconds by the Earth's rotation but digital cameras have two big advantages. The first is that the sensor is likely to be more sensitive than film so you should get more comet detail with the same exposure. The second advantage is that you can take multiple consecutive short exposures of the sky and then stack them into a single image shifting each one to counteract the sky's rotation. In this way you can simulate a long exposure without having access to a driven camera mount. If you do have a drive then you can do longer exposures, anyway, but it is sometimes advantageous to keep the exposure short so that the brightest part of the coma is not saturated. In this case you would be adding together multiple frames to increase the dynamic range of the output image. This can then be processed with a non-linear stretch and will show detail in both the coma and the tail. Such a result is very difficult to achieve with film.

Remember that digital cameras have no cooling so exposures of many seconds will be marred by thermal noise. This can be overcome if you ensure that the exposure is kept short enough that none of the pixels in the image is saturated. When your exposure is complete take a dark frame of exactly the same exposure length and then subtract this from the original image to remove the thermal noise. In fact some digital cameras do this automatically for long exposures. If you select a

10 second exposure and it takes 20 seconds for the image to appear this the most likely cause. Dark frames and other tricks will be covered in more detail in Chapter 8.

Special Projects

Now you have mastered the techniques of wide-field comet imaging you may be wondering what to do next. Comets are fascinating objects and there are always interesting projects that can be done when a reasonably bright one is in the sky.

Have you ever used a stereo viewer to look at 3-D photographs? If you have you will know how effective this technique can be and how simple it is. When we look at something each eye is viewing the scene from a slightly different position and our brain processes these images to give us depth perception. In a stereo camera our eyes are replaced by two lenses and the camera produces two photographs of the subject from slightly different positions. The stereo viewer allows us to view the two photographs, one with each eye, and the illusion of depth is created by our brain.

A fun project is to take a stereo image of a comet. Of course it is not possible to take a true 3-D image since most comets are so far away that you would have to place your lenses further apart than the diameter of the Earth to get a suitable effect! What you can do is take two images of the comet separated by enough time that it has moved relative to the star background. A few minutes or hours between exposures will suffice depending on the focal length you are using and the speed the comet is moving relative to the stars. Align the camera or print the image so that the comet is moving horizontally and it is furthest to the right in the left hand frame. You can now view the image using a stereo viewer and you should see the comet floating in front of the background stars. If you don't have a stereo viewer you can get the same effect with no equipment at all as long as you are prepared to do some visual gymnastics (Figure 5.14).

Figure 5.14. These two pictures of C/1996 B2 Hyakutake were taken two hours apart on 1996 March 25 and they are printed as a stereogram. To view them in stereo without special equipment you can use a technique called "free-fusion". Hold a pencil between you and the picture. Close your left eye and move the pencil back and forth until its tip lines up with a bright star in the left hand image. Then swap eyes and move the pencil so that it lines up with the same bright star in the right hand image. Swap back and forth between eyes until you are happy that the pencil lines up with the star in both cases. Now concentrate on the tip of the pencil and you should see the two comet images merge together in the background. Move the pencil away and re-focus on the images *without* moving your eyes (this is the hard part). If you succeed you should see the comet floating above the star background.

The stereo image is interesting but what can you do if you are interested in producing science? In very bright comets the gas tail can show a great deal of detail even in wide field images. The tail is effectively acting as a windsock in the gusty solar wind and until the advent of space probes comet tails were the best way of measuring solar wind features. Sharp changes in the solar wind can force the tail to break away from the coma and these *disconnection events* can be particularly interesting and beautiful (Figure 5.15) especially when imaged at high resolution. Taking a sequence of photos of a bright comet provides an excellent record of its development as long as you remember to record all of the important details for each exposure. At a minimum you should record the start and end times of the exposure, the focal length and f/ratio setting and the sky conditions.

Cometary dust tails can also show considerable detail but this is often difficult to pick up with short focal length lenses. Striations in the tail caused by an episodic

Figure 5.15. These three photographs of C/1996 B2 Hyakutake have been printed to the same scale and orientation to show how the tail developed on the morning of 1996 March 25. At the time these images were taken two disconnection features are visible in the tail. A & B are 85 mm lens + TP2415, C is 55 mm lens + TP2415.
Nick James, Martin Mobberley and Glyn Marsh.

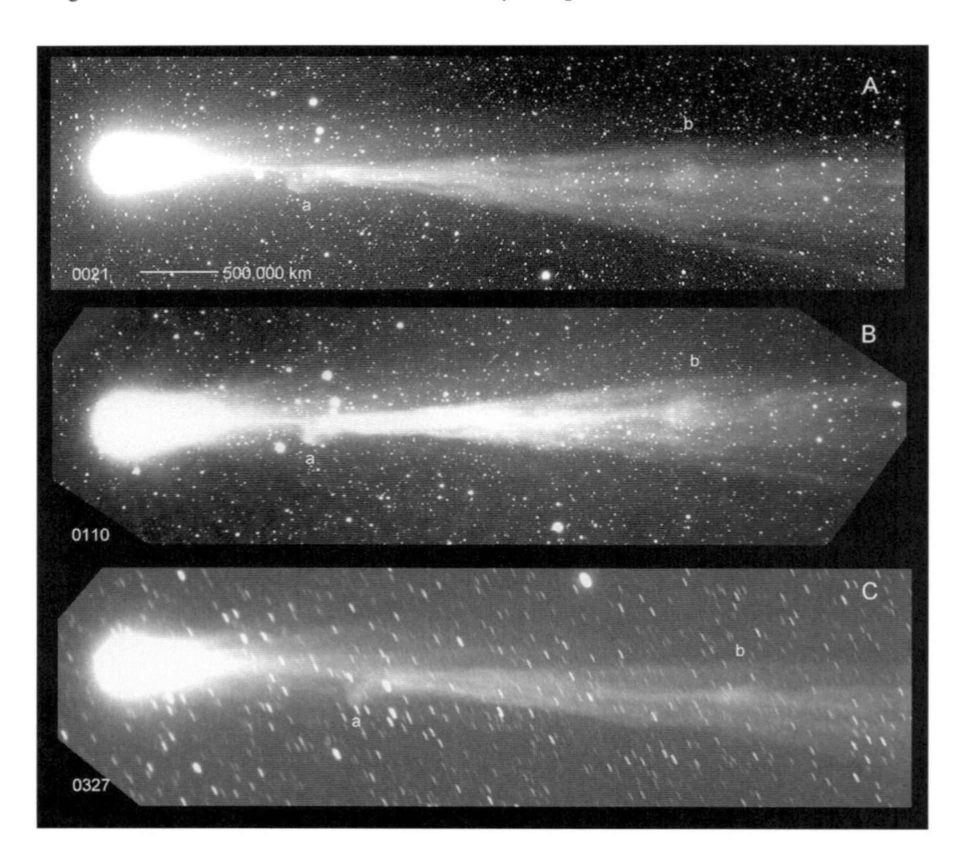

release of dust from the nucleus can be prominent. These *synchronic bands* were well developed in the dust tail of Hale–Bopp (Chapter 10) and they give us some clues as to how the dust is being released from the rotating nucleus.

The evolution of tail and coma features can best be shown if you make a movie from your comet images. As you will have seen comets are very dynamic objects. Not only do they move against the background stars but their physical appearance changes from night to night and sometimes even from hour to hour. This dynamic behaviour is shown up very well in time-lapse movies where the passage of hours or days is compressed into seconds. In the past these movies were filmed using special cine cameras which could be programmed to take long exposures on successive frames of film separated by fixed intervals. Special equipment was required to ensure that the comet was correctly registered in each frame since the time lapse effect is ruined if the comet jumps around during the movie.

You will be pleased to know that time-lapse movies can now be created using conventional still cameras and liberal amounts of digital post-processing. A good project would be to image the development of a comet's tail from night to night. If you want to try this make sure that you stick to the same settings (exposure, aperture, etc.) on each night and keep the comet in roughly the same place on the detector. This sequence might last for weeks so it is important to plan your exposures in advance. Once you have the frames you should process them so that they all have the same sky background level and align them so that the comet's head is in the same position on each frame. The individual frames can then be combined into a movie using suitable computer software. I will cover this procedure in detail in Chapter 8. If you are lucky enough to get a good sequence of clear nights I'm sure that you will be impressed with the results. If you record the details of each exposure the sequence will also be very useful scientifically. You will find some good examples on the CD-ROM which is supplied with this book.

One of the best movies of a comet ever made was obtained by the Canadian Peter Ceravolo and his colleagues. They took 900 images of Hyakutake in the week just before close approach using medium format Fuji SG800 film and 440 mm FL, f/2.3 optics giving a field of view approximately 6° square. Each negative was scanned and they used specially written software to align each frame and combine them into a movie. If you are interested in making comet movies of your own I suggest that you view Ceravolo's video for inspiration (see the Appendix). It shows just what you can achieve with a good camera, the right software and tremendous dedication!

In the meantime we need to consider how computers can assist the comet observer in other ways.

Chapter 6
Comets and Computers

One of the things that makes comets such fascinating objects is that the very best ones usually come as a complete surprise. Most of the brightest comets are in very long, thin orbits which have periods of thousands of years. They are usually discovered a few months or, exceptionally, years before they reach their best and most of them behave in a highly unpredictable manner. In fact, probably the only thing that can be predicted about a comet is its position as it moves across the sky. The computation of cometary orbits used to be a highly specialised activity but the rapid development of computer hardware and software means that it is now within the capabilities of the interested amateur.

Orbit computation is only one of the many areas where computers have revolutionised cometary astronomy. You can use a computer to predict where a comet will appear in the sky tonight or where it would appear thousands of years in the past or future. Your computer can then help you to point your telescope at it, guide on it and, with a CCD camera, the computer can collect and store your digital images. It can then process those images, extract data from them and format that data in a way that makes it easy to submit the observations to national organisations – by e-mail of course. Some amateurs have even discovered comets without ever using a telescope. They have used images from the SOHO spacecraft downloaded from the web and processed in their home computers to find comets that the professionals have missed. In short, computers are now almost as important as telescopes when it comes to studying comets and many would say that they have become an indispensable tool in the armoury of the comet observer.

We will start our discussion of these applications by looking at how comets move around the Solar System. You can't observe a comet without knowing where it is and we can only predict where a comet will be if we have an accurate knowledge of its orbit.

Simple Orbits

As with all of the other objects in the Solar System, comets are moving under the gravitational influence of the Sun and planets. You probably remember the famous, but apocryphal, story of Sir Isaac Newton and his apple. The story goes that the sight of the apple falling to the ground encouraged Newton to develop his universal theory of gravitation. This states that the force between any two objects in the universe is proportional to the product of their masses divided by the square of the distance between them. Mathematically this is written as:

$$F = G\, M_1 M_2 / d^2$$

where M_1 and M_2 are the masses of the two objects involved, d is the distance between them and G is a fundamental number called the *Gravitational Constant.* Every object in the Solar System exerts a force on every other object but since the mass of the Sun is so much larger than even the largest planet it dominates orbital motion in the Solar System. In fact, in most cases, we can completely ignore the forces due to everything else and say that the comet is orbiting around the Sun in what is called a two-body orbit. Two body orbital motion is relatively simple to analyse since the only force acting on the comet is the gravitational attraction given by Newton's equation and this force is directed towards the centre of the Sun.

In two-body motion the orbital path is closed, that is after one orbit the comet comes back to exactly the same place relative to the Sun as it started. The simple equations of two-body motion lead directly to the laws of orbital motion put forward by Kepler which were introduced in Chapter 1. Putting aside a few complications that I will explain later, the orbital history of the comet for all time is described by only seven numbers: at a given time t, we need to know the position in three dimensional space (given by the triplet x, y, z) and the velocity (given by the triplet x', y', z'). No further information is required so if you know the comet's position in space and its velocity at any time you can predict the position and velocity at any other time in the future or the past using a simple set of equations.

Whenever a comet is discovered you will usually see the *orbital elements* published in astronomy magazines or on the web. An initial orbit can be computed when we have three accurately timed *astrometric* positions separated by a few days. Astrometry is the process of measuring the position of a comet relative to the background stars. Good astrometry is fundamental to our understanding of comets and their orbits and it forms the basis for all other observations. It is an area where the amateur observer excels and I will have much more to say about astrometry in Chapter 9. Quite often the initial orbit will allow observers to find pre-discovery images of the comet maybe days or weeks beforehand. In this way our knowledge of the orbit rapidly improves and in most cases the initial orbit is sufficiently accurate to predict the position of the comet many weeks or months into the future. However, this is not always the case and, particularly if the comet is far away and slow moving, it can take quite a while before we are sure of the orbit. Comet Hale–Bopp in 1995 was a particularly good example of this since it was discovered when it was far away and moving almost straight towards us.

Rather than giving the position and velocity at a particular time the orbital elements that you will see in magazines describe the shape and orientation of the comet's orbit and its position along that orbit. This makes the elements much easier for us humans to understand but it is a relatively simple procedure to transform between this type of elements and the position/velocity pair. In fact computers usually work with the latter but will accept the former for convenience. Some example elements are shown in Table 6.1. These are for the famous short-period comet, P/Halley. How do these elements relate to the shape of orbit and how do they describe where the comet will appear in the sky?

The shape of the orbit is described by the eccentricity, e, and all comets follow orbits which are known as *conic sections* (Figure 6.1). These range from near-circular orbits where e is close to zero, through longer and longer ellipses as e increases, to just less than one

Table 6.1. Orbital elements for comet 1P/Halley

Epoch: 1986 Feb 19.0 (JDT = 2446480.5)		
T = 1986 Feb 9.4590 TT	$\omega = 111^\circ.8657$	J2000.0
$e = 0.967277$	$\Omega = 58^\circ.8601$	J2000.0
$q = 0.587104$ AU	$i = 162^\circ.2422$	J2000.0
$A_1 = +0.04$, $A_2 = +0.0155$		

to parabolas where e is exactly one. Some comets are actually found to be in hyperbolic orbits (e > 1) but these are rare. Comets in truly parabolic and hyperbolic orbits have infinite orbital periods and so they visit the inner Solar System only once in their lives. Many new comets are initially assumed to be in a parabolic orbit but are later found to be in a highly eccentric ellipse. Comets that are in true hyperbolic orbits will leave the Solar System never to return and they have usually been ejected following a close planetary encounter although there is speculation that some were visitors from interstellar space.

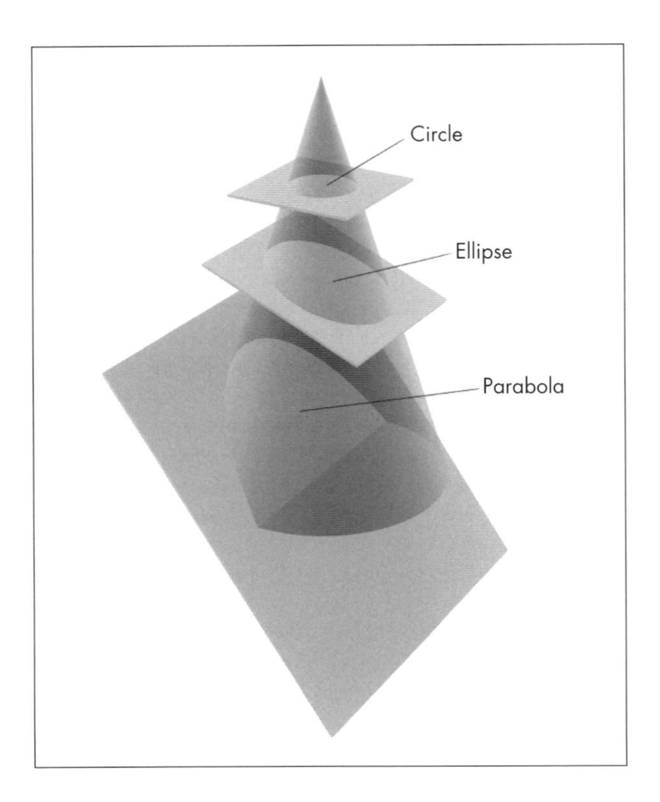

Figure 6.1 The four types of comet orbits can be formed by slicing a cone. Three of the possibilities are shown here. A circle is formed by slicing through the cone horizontally. An ellipse results from an inclined slice with the eccentricity increasing as the slice gets steeper. A slice which passes through the centre of the base of the cone forms a parabola and even steeper slices form hyperbolas (not shown).

Kepler's first law states that the Sun is at one focus of the ellipse (Figure 1.11 on page 13). The value q gives us the minimum, or *perihelion*, distance from the Sun to the comet. This distance is always measured in astronomical units (AU) where one AU is the average distance from the Earth to the Sun, about 150 million km. Normally, for a comet to be spectacular, q should be less than 1 AU so that at perihelion the comet is within the orbit of the Earth. Comets often brighten as the inverse fourth, fifth or sixth power of their distance from the Sun so comets with small values of q can be very spectacular indeed. Some comets, called *Sungrazers*, have values of q less than 0.01 AU and such comets can be extremely bright at perihelion although at their best they are necessarily close to the Sun and so are only seen against a bright sky. Many Sungrazers have been discovered in recent years by observers using the coronagraph onboard the SOHO spacecraft . This has even recorded comets which have q values so small that they actually fall into the Sun.

The elements e and q tell us the shape and size of the orbit but we also need to describe its orientation in space. Before we can do this we need a suitable coordinate system. The system used is called the *equatorial coordinate system* and it is based on two planes. The Earth's orbit defines one of the reference planes and the second plane is given by an extension of the Earth's equator. These two planes are inclined to one another by an angle of about $23\frac{1}{2}^\circ$ which is called the *obliquity of the ecliptic*. Since the planes are inclined to one another they intersect along a line as shown in Figure 6.2. The Earth moves around the Sun in a slightly elliptical orbit but for the purposes of this coordinate system we assume that the Earth's orbit is circular with a period of one year. Twice a year at the equinoxes this idealised Earth passes through the line of intersection of the two reference planes. At the spring equinox a line joining the Earth to the Sun passes through both planes and points to a direction in space known as the First Point of Aries. This direction is normally given the symbol ♈.

We can now return to describing the comet's orbit. The angle between the plane of the Earth's orbit and the plane of the comet's orbit is called the *inclination* and given the symbol i. By convention the inclination is less than 90° if the comet is moving around its orbit in the same direction as the Earth (prograde motion) and greater than 90° if the comet is moving in the opposite

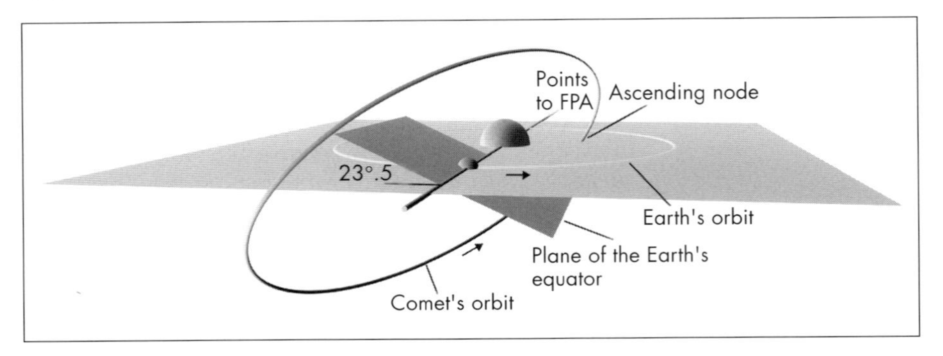

Figure 6.2 The intersection of the two planes defined by the Earth's orbit and its equator are used as the reference direction in the equatorial system. Comets move in orbits which are inclined relative to the Earth's orbit and so they pass from below to above the plane at the ascending node.

direction (retrograde motion). Halley's comet is moving in the opposite direction to the Earth and hence its inclination is given as 162°.2.

For any inclination other than 0° and 180° the comet will, at some point move from below the Earth's orbital plane to above it. The point at which is does this is called the *ascending node* and the angle between the First Point of Aries and the comet's ascending node measured along the plane of the Earth's orbit in the direction of the Earth's motion is called the *longitude of the ascending node* and it is given the symbol ☊. The final angular element is a measure of the angle between the ascending node and the line joining the comet to the Sun when the comet is at perihelion. This angle, measured in the plane of the comet's orbit in the direction of the comet's motion is called the argument of perihelion and it is given the symbol ω.

The final element in our set is the time at which the comet reaches perihelion, T. So, therefore we have six elements, T, q, e, i, Ω and ω which completely describe the two-body orbit for all time. You might remember that I said that seven parameters were needed. Actually the element T combines two elements into one since it defines the time of perihelion and this tells us both the time and the position of the comet in its orbit.

While q defines the perihelion distance, you can quickly determine the aphelion, Q, or furthest out, distance of the comet using:

$$Q = q \times (1 + e)/(1 - e)$$

You may occasionally see elements published in a slightly different format. These are often used to describe the motion of asteroidal objects which move in near-circular orbits, very different to the highly eccentric orbits of most comets. In this alternative scheme the perihelion distance, q, is replaced by the semi-major axis of the orbit, a. The time of perihelion, T, is replaced by a pair of elements representing the *epoch* of the orbit and the mean anomaly, M, or angular distance around the orbit, at the time of the epoch. You can transform between a and q as follows:

$$a = q/(1 - e)$$

You can demonstrate why the "asteroidal" elements are not suitable for highly eccentric comets by putting $e = 1$ in the above equation. The semi-major axis becomes infinite no matter how small q is. While it is not one of the elements the period of a comet's orbit, P, depends on the semi-major axis alone. In fact

$$P = a^{3/2}$$

where P is measured in years and a is in AU.

So far, this is all very neat and tidy but you should be aware that there are a few complications that can sometimes be confusing, even to experienced observers. The first complication arises because our planet does not have a constant spin axis. Rather like a child's spinning top the Earth's rotation axis precesses in space taking approximately 26,000 years to complete a full wobble. If you remember, one of our reference planes was an extension of the Earth's equator. Precession and other effects are responsible for a continual change in the direction of the intersection of the Earth's equator with the plane of its orbit and this directly affects the equatorial coordinate system.

Any position quoted in equatorial coordinates must include a reference date which tells us which equator and equinox is being used. Our comet elements are said to be referred to the equator and equinox of such and such a date. Up to the 1980s the reference date was just before the start of the year 1950 (called B1950.0) but it is now just after the start of 2000 (called J2000.0). The orbital elements that you will see published in recent magazines should be exclusively J2000.0 but you will encounter B1950.0 elements in some older publications. Computers can transform between these different systems without any problem but it is important that they know which one is being used!

The second complication that you should be aware of involves the measurement of time. Astronomers have evolved many different timescales which are used in measuring different things. The timescale used for orbital elements and ephemerides is a uniform timescale based on atomic clocks called Terrestrial Time (TT). Our normal civil time is called Coordinated Universal Time (UTC), but it is not suitable for orbital calculations since it is related to the spin rate of the Earth and this is not uniform. TT differs from UTC by an amount called ΔT. To keep UTC synchronised with the rotation of the Earth it is necessary to introduce "leap seconds" every now and again. In 2001 the difference TT minus UTC was 66 seconds and this has been gradually increasing at a rate of about a second per year for the last few years but future changes are unpredictable. The value of ΔT is only important in astrometric studies although even here observation times should always be reported in UTC. For really accurate work though you should remember that the time of perihelion, T, and the times associated with all positions predicted by orbital elements will be in TT, not UTC.

Perturbations and Comet Families

Up to now we have ignored the effect of other Solar System objects on the orbit of our comet. This is a reasonable thing to do in the short term but in the longer term all of the small forces resulting from other objects in the Solar System will accumulate and perturb the comet away from the orbit predicted by simple two-body motion. When more than two bodies are involved in orbital motion there is no neat, general solution corresponding to the seven orbital elements but in most cases the perturbing effects are still small and so we can still use the familiar orbital elements with a small addition. At any particular time we assume that the comet is following a closed orbit defined by a set of orbital elements but we accept that this orbit changes with time. The elements are called *osculating elements* and they come with an extra piece of information called the *epoch* which defines the time at which the elements correctly describe the closed orbit. You can use osculating elements to predict a comet's position at any time in the future or the past but the expected errors in position will grow as you move away from the epoch date.

Since there is no general set of equations that we can use for three or more bodies in orbital motion the only way of accurately predicting future positions is to effectively construct a mathematical model of the Solar System and gradually move forward or backwards in time calculating all of the perturbations at each step. This process of *numerical integration* used to be extremely time consuming but there are now many computer programs which do all of the hard work for you. All you need to do is provide the computer with the orbital elements for some given epoch and it will integrate the orbit forwards or backwards in time at the touch of a button. The computer will be able to give you the osculating elements, and hence the comet's position, for any date with an accuracy which is dependent on the number of perturbing objects used and the step size of the integration. An example orbit integrator is John Chambers' *Mercury* which is available as FORTRAN source code from his website (see the Appendix). This compiles without any problems on Unix systems. More examples are present on the CD-ROM.

Integrating an orbit forwards or backwards in time can produce some very interesting results. Some comets have seen dramatic changes in their orbits at some point in the past due to close encounters with planets. Numerical integration is the only way to trace the path of a comet back in time so that we can investigate where it originally came from. An example integration is shown in Figure 6.3.

In Chapter 2 we mentioned the Oort Cloud. This is the vast reservoir of deep-frozen comets which was originally proposed by Jan Oort in 1950. The primary evidence for this cloud comes from the integration of comet orbits. The results show that when the current orbits of long-period comets are integrated backwards in time most of them seem to have started with aphelia somewhere between 50,000 and 150,000 AU from the Sun. What's more the orbits of these long period comets seem to be randomly oriented in space. This is powerful evidence for a spherical shell of comets which stretches about a quarter of the way to the nearest star.

It used to be thought that all comets came from the Oort cloud but more recent orbital calculations showed that it could not sustain the population of observed short-period comets. Rather than having random orientations, short-period comets tend to move in low inclination, prograde orbits and there are far too many of them to have been captured from long-period orbits

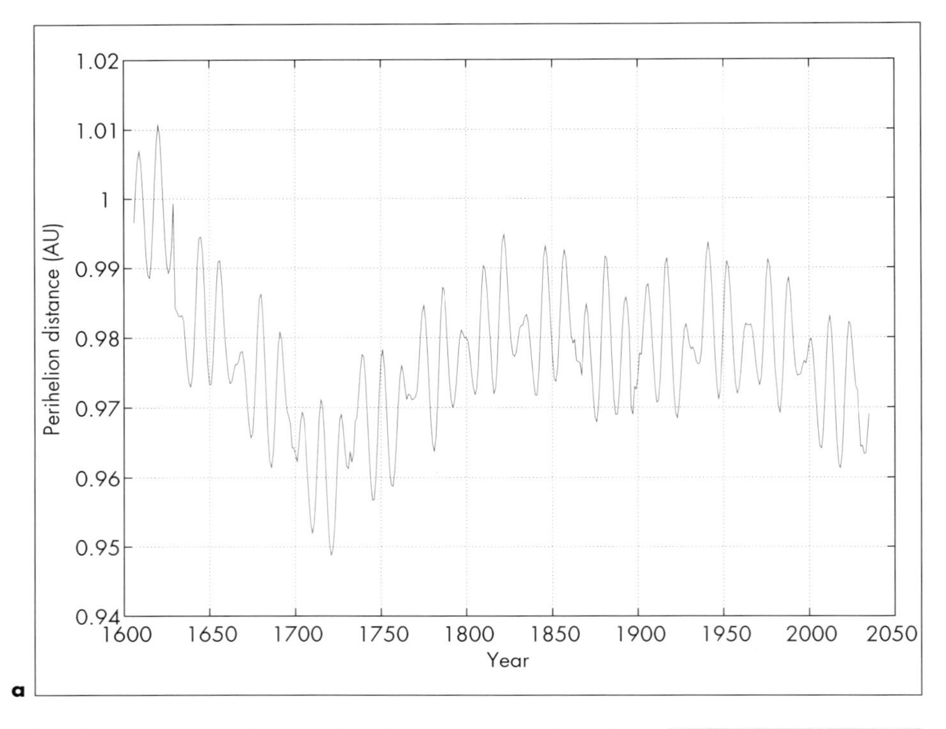

Figure 6.3. These plots show the evolution of perihelion distance **a** and longitude of perihelion **b** for Comet 55P/Tempel–Tuttle. This comet has an orbital period of 33.2 years. The perihelion distance shows a 11.9 year modulation corresponding to the orbital period of Jupiter. The steps in the longitude of perihelion occur at 166 year intervals when the comet has a particularly close approach to Jupiter. These close approaches occur regularly since the comet completes five orbits in exactly the same time that it takes Jupiter to complete 14.

by planetary perturbation. People began to realise that there must be a reservoir that was much closer to the inner Solar System and it just so happened that such a reservoir had been discussed even before the Oort cloud was postulated.

In 1943 Kenneth Edgeworth published a paper in the *Journal of the British Astronomical Association*. In this paper he wrote about something similar to the short-period comet reservoir that was required. This was a full seven years before Oort proposed his cloud but it didn't attract much attention. In 1951 Gerald Kuiper proposed a similar thing but it was not until 1992 when professional telescopes started to detect huge dormant comets in the frigid space beyond the orbit of Neptune. The belt has become known as the Edgeworth-Kuiper Belt (EKB) although there is some debate as to whether Edgeworth or Kuiper had really predicted what we can now observe. Many astronomers prefer to call these dormant comets *Trans-Neptunian Objects* (TNOs). Almost 500 TNOs are now known and they have aphelia of up to 200 AU with inclinations which are mostly near to the ecliptic plane (Figure 6.4). The brightest TNOs reach magnitude 20 and so they are beyond the range of most amateur observers. There has

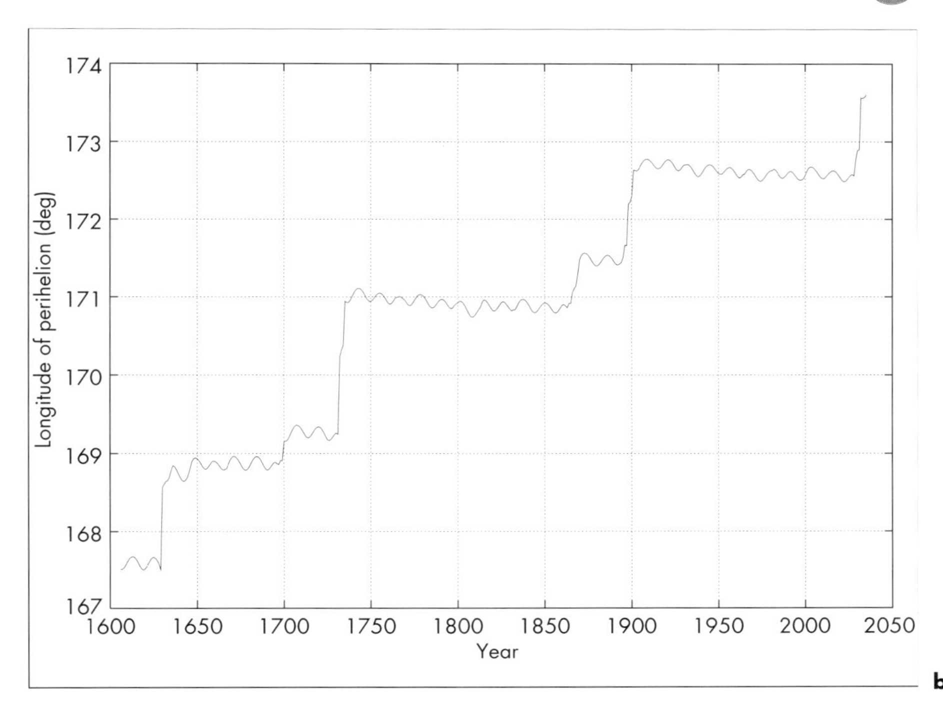

Figure 6.3b.

been some debate as to whether the outer planet Pluto and its satellite Charon are themselves TNOs. If they are then it is possible to see a huge dormant comet in the outer Solar System with a modest telescope since Pluto reaches magnitude 14 at opposition.

Another interesting aspect revealed by the integration of orbits is that certain comets seem to belong to families. When the orbits of these comets are integrated backwards in time it seems that they all converge on a single object, as if a huge comet had broken up into multiple objects at some time in the past. One of the largest families is known as the *Kreutz group* after Heinrich Kreutz. Around 1890 he suggested that many of the sungrazing comets appeared to be in very similar orbits. He proposed that there was once a massive comet that broke up at perihelion and which was the parent of all of these objects. Brian Marsden used the techniques of orbit integration to show that the original comet probably split into two many thousands of years ago. The two parts were themselves disrupted at perihelion passages in 372 BC and 1106 AD respectively to form two subgroups called Kreutz I and Kreutz II. The original comet must have been huge with a nucleus

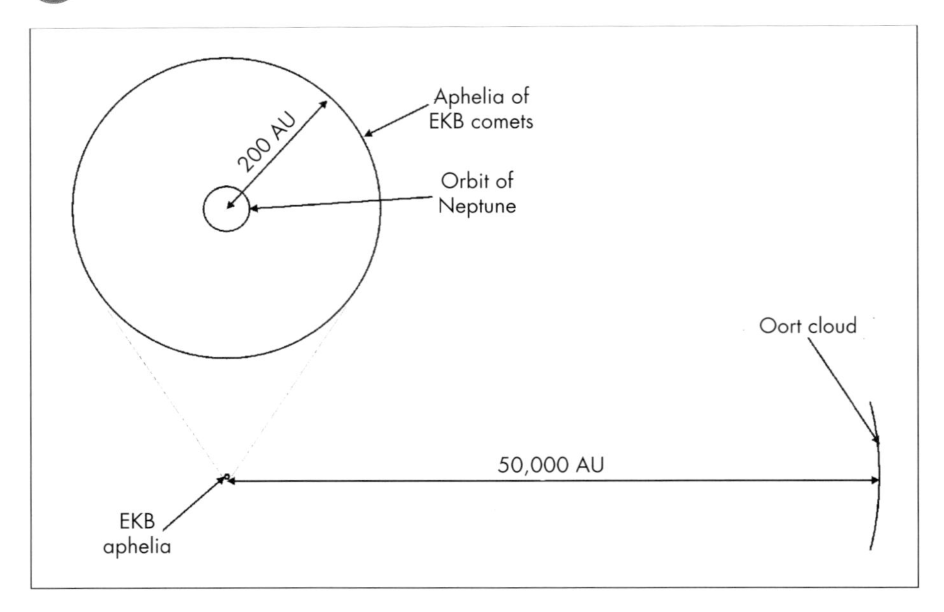

Figure 6.4 The relative dimensions of the Edgeworth–Kuiper belt and the Oort cloud are shown in these two diagrams.

perhaps 120 km in diameter. Other families exist and it is only by integrating current orbits backwards using computers that we can hope to understand the history of these objects.

Sometimes, rather than continually changing a comet's orbit, gravitational perturbations will lock a comet into what is called a *resonance*. The comet might, for instance, have an orbital period in which it performs three revolutions in exactly the same time that Jupiter takes to go twice around the Sun. After three revolutions the comet returns to exactly the same position with respect to Jupiter and so the pattern of gravitational perturbations repeats. Such an orbit is called resonant and this particular case would be described as a 3:2 resonance. A number of periodic comets are known to be in resonant orbits with Jupiter, one of the most interesting being the progenitor of the Leonid meteor shower, Comet 55P/Tempel–Tuttle (Figures 6.3 and 6.5) whose 33 year orbit is locked in a 5:14 resonance with the giant planet.

The material evaporated from a comet trails out along the orbit and if the Earth should encounter it we will see a display of meteors. The Leonids which are seen each November have a greatly enhanced level of activity near to times when the parent comet is at perihelion. Most recently Leonid storms were seen in 1998, 1999 and 2001 following Tempel–Tuttle's perihelion passage in February

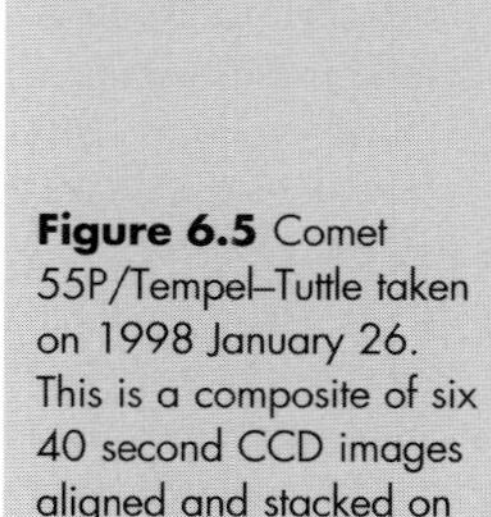

Figure 6.5 Comet 55P/Tempel–Tuttle taken on 1998 January 26. This is a composite of six 40 second CCD images aligned and stacked on the comet.

1998. It is only recently that accurate predictions of these meteor storms has been made possible by people such as Rob McNaught and David Asher. They have used orbit integration calculations to model perturbation effects on all of the particles in this stream of cometary debris.

The orbital history and future path of a comet is a fascinating area of study which you can carry out on a relatively modest PC. Computers are ideal for simulating the motion of objects in the Solar System over millions of years. They never get bored and you can take the credit if they find anything interesting! This work can also identify comets as members of certain families or make predictions of meteor activity. It is ideal work for the comet enthusiast who doesn't have access to good skies or a telescope but who wants to make an important contribution to cometary science. The source code for orbit integrator programs is available on the Internet and we have listed some examples in the Appendix. A good starting point for current orbital elements is the Catalog of Cometary Orbits published by the Minor Planet Center (see the Appendix). It contains orbits for over 1300 comets but there are other sources freely available on the web.

Natural Rockets

A comet does not always stick to the path that we would expect, even after we have taken account of all of the various gravitational effects since other perturbing forces are present. These forces are due to the "rocket effects" that we mentioned in Chapter 2. Of all the objects in the Solar System only the comets have jets from their surfaces which make them act like natural rockets. This phenomenon provides us with a tool with which we can gain some insight into the physics and nature of cometary nuclei.

The non-gravitational forces (NGFs) on the nucleus are usually modelled using two parameters, A_1 and A_2. The actual force on the nucleus is dependent on how far the comet is from the Sun since the rocket effects are directly proportional to the rate of ejection and this increases rapidly as the nucleus is heated. The parameter A_1 represents a force which acts radially away from the Sun. If the nucleus is not rotating we would expect all of the jets to be on the sunward side of the nucleus and so the resulting rocket force would tend to push the comet radially away from the Sun. In many cases however it is necessary to add a second force, A_2, which acts tangentially along the orbit. This force can be explained if we assume that the nucleus is rotating since the jets will be displaced with respect to the Sun–comet line. It is only through accurate astrometry and subsequent orbit computation that we can derive the non-gravitational force parameters. Theorists can then use A_1 and A_2 to improve our understanding of the physical nature of the comet.

Even though NGFs are small they can have substantial effects over long periods. The original calculations for the size of the Oort cloud suggested a radius of 100,000 AU but this neglected the NGFs. If NGFs are taken into account the radius shrinks to around 50,000 AU. A substantial change for such a small force.

Ephemerides and Observability

If you don't plan your observing carefully you may find that the reward for getting up early to image a comet is that it is hidden behind a neighbour's tree, or the sky is too bright due to twilight or a nearby Moon. Even if you can get your telescope aligned on the comet you may find that after spending hours out in the cold and damp you manage to get a picture with the coma on the wrong side of the field and the tail streaming out of the frame. All of this can be avoided if you carry out some simple calculations long before you go outside. Planning is especially important if you are travelling long distances to see the comet.

If you are keen to observe new comets shortly after they are discovered you will need to subscribe to one of the alert services run by professional or amateur bodies (see the Appendix). Many of these provide paper circulars via conventional mail but most people now receive alerts via e-mail. The first thing I do when I hear of a new comet is to check how it is likely to appear in my sky. This is easy to do once you know the orbital elements since they will allow you to predict where a comet will be in space at any time. It is then straightforward to predict where it will be seen from a particular location on the Earth. You will quite often find a list of positions called an *ephemeris* for a particular comet printed in magazines but it is usually better to produce your own since you can tailor it to your observing site and the time that you are likely to be making the observation. In an ephemeris the comet's position (normally in Right Ascension and Declination) is listed for each date of interest but it might also contain other information such as the distance of the comet from the Earth and Sun, the expected magnitude and the motion of the comet on the sky.

The first thing to consider when planning to observe a comet is the most obvious. When will it be above the horizon in a reasonably dark sky? What constitutes reasonably dark depends on the brightness of the comet but Nautical Twilight (the time when the sun is 12° below your horizon) is normally a good limit. It is also important to check whether the Moon is above the horizon since this natural source of light pollution can drown all but the brightest of comets. The next thing to consider is will the comet be above the local horizon at your favourite observing site? You may well have a rather limited horizon and houses and trees always conspire to be in the most inconvenient position possible. It is best to know about this before you get up at 4am and find that the view to the comet is blocked! Unlike other astronomical objects you can't just wait for the comet to move into an unobstructed area of sky since the Sun is usually not very far away. If the trees are yours it is possible to trim them when a particularly bright comet comes along but the situation is rather more difficult if they are your neighbour's! Comets are at their best when near to the Sun so they usually appear low in the west in the evening or low in the east in the morning. The ideal observing site would therefore have clear eastern and western horizons. It is very rare to see a bright comet at a large elongation from the Sun in a dark sky, although fainter comets can often be more conveniently positioned.

There are many good computer programs that will accept cometary orbital elements and will then produce an ephemeris or will draw the track of the comet against the background stars. In many cases you will not even need to enter the elements yourself since they can be downloaded in the correct format from the Central Bureau's website. A particularly useful output of the planetarium programs is a plot of where the comet will appear in the morning or evening sky at Nautical Twilight. This will tell you whether you will be able to see the comet from your normal site or whether you need to go somewhere else. An example from a program called GUIDE are shown in Figure 6.6.

There are many computer planetarium programs on the market. I have used GUIDE from Project Pluto but most of the ones you will see advertised can accept orbital elements and produce plots for you. See the resources section for further suggested programs. If you don't have any of the common computer planetarium programs don't worry. There are now some excellent

Figure 6.6 A screenshot from Guide 7 showing the position of C/1995 O1 on 1997 March 28. Compare with the photograph shown in Figure 5.3. Courtesy project PLUTO.

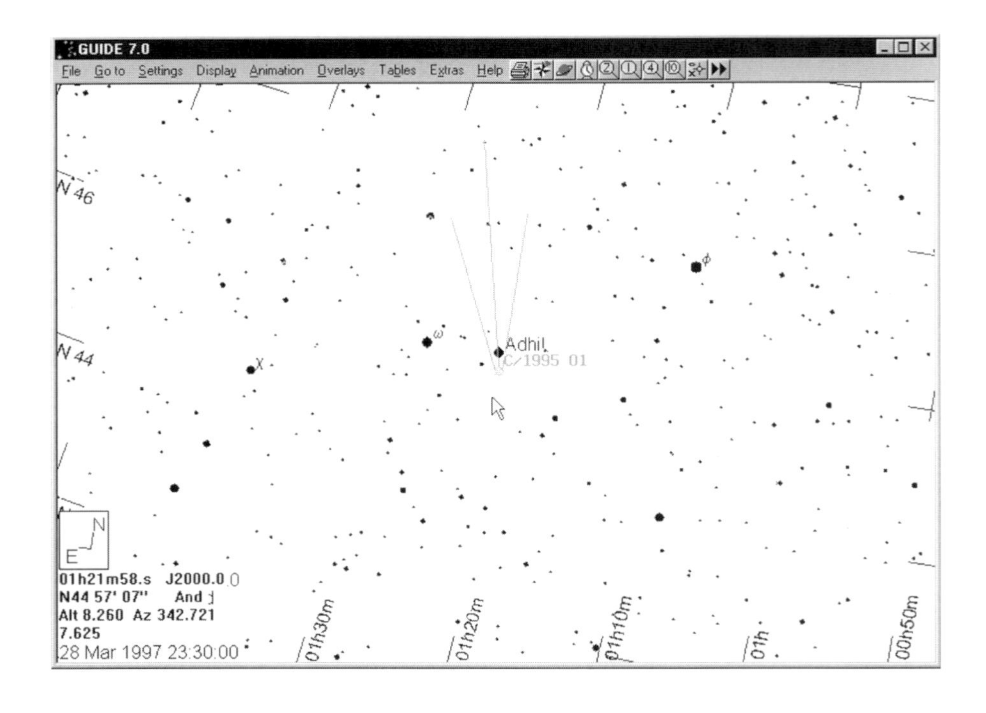

websites which will generate the charts for you, tailored to your observing site and conditions. Have a look at the Heavens Above site for a fine example of what can be done. JPL also have an excellent website called Horizons which will compute extremely accurate ephemerides of Solar System bodies on demand. This is the site that I always use. An example ephemeris from this website is shown in Table 6.2.

There are some very useful programs which will produce "observability diagrams" from ephemerides. One such is Graphdark by Richard Fleet (Figure 6.7). This gives a rapid graphical overview of the best times to observe a comet by displaying twilight times, rising and setting times for the Moon and various other important parameters on a single plot.

Once you have established when and where the comet will appear the next thing to consider is how bright it is expected to be and the direction of any possible tail. The brightness can be obtained from the ephemeris but it is often not so easy to determine the tail direction. The easiest approach is to look at recent observations which report the tail *position angle* (PA) (Figure 4.6 on page 83). If this fails it is usually possible to determine a rough tail PA by projecting a line away from the Sun onto the sky. This is not an infallible approach but you should find that your favourite computer planetarium program can display this information.

Many of the planetarium programs will make estimates of how bright the comet will be in addition to telling you where it will be. These estimates are based on magnitude parameters supplied with the elements but they should be treated with great caution. While we can predict positions from the elements very accurately the magnitude of a comet is subject to many uncertainties. The usual equation for a comet's apparent magnitude is:

$$m_1 = H_1 + 5.0 \log_{10}(\Delta) + K_1 \log_{10}(r)$$

where H_1 and K_1 are magnitude constants which depend on the size and the activity of the comet and Δ and r are the distance of the comet from the Earth and Sun respectively (in AU). The 5.0 $\log_{10}\Delta$ term is simply a statement of the inverse square law in terms of magnitudes. All other things being equal a comet which is twice as far from the Earth will be one fourth as bright (or 1.5 magnitudes fainter). The $K_1 \log_{10} r$ term tells us how much more intrinsically bright the comet

Table 6.2. Ephemeris from JPL horizons for C/2002 C1 Ikeya-Zhang in early 2002. The columns show the date, position (RA and Dec), the comet's motion in RA and Dec, the total magnitude and nuclear magnitude and the current constellation.

Date__(UT)__HR:MN	R.A._(ICRF/J2000.0)_DEC	dRA*cosD	d(DEC)/dt	T-mag	N-mag	Cnst
2002-Feb-09 00:00	00 22 01.02 –13 42 45.7	66.74	83.44	8.33	n.a.	Cet
2002-Feb-10 00:00	00 23 50.57 –13 09 07.3	67.79	85.06	8.23	n.a.	Cet
2002-Feb-11 00:00	00 25 41.57 –12 34 49.1	68.83	86.75	8.14	n.a.	Cet
2002-Feb-12 00:00	00 27 33.98 –11 59 49.5	69.85	88.49	8.04	n.a.	Cet
2002-Feb-13 00:00	00 29 27.79 –11 24 07.1	70.86	90.31	7.94	n.a.	Cet
2002-Feb-14 00:00	00 31 22.96 –10 47 40.2	71.84	92.19	7.84	n.a.	Cet
2002-Feb-15 00:00	00 33 19.45 –10 10 27.1	72.80	94.14	7.73	n.a.	Cet
2002-Feb-16 00:00	00 35 17.22 –09 32 26.2	73.72	96.16	7.62	n.a.	Cet
2002-Feb-17 00:00	00 37 16.22 –08 53 35.5	74.61	98.26	7.52	n.a.	Cet
2002-Feb-18 00:00	00 39 16.39 –08 13 53.3	75.46	100.45	7.41	n.a.	Cet
2002-Feb-19 00:00	00 41 17.67 –07 33 17.5	76.27	102.71	7.30	n.a.	Cet
2002-Feb-20 00:00	00 43 19.99 –06 51 46.1	77.01	105.06	7.18	n.a.	Cet
2002-Feb-21 00:00	00 45 23.25 –06 09 17.1	77.69	107.50	7.07	n.a.	Cet
2002-Feb-22 00:00	00 47 27.36 –05 25 48.2	78.30	110.03	6.95	n.a.	Cet
2002-Feb-23 00:00	00 49 32.21 –04 41 17.2	78.83	112.65	6.83	n.a.	Cet
2002-Feb-24 00:00	00 51 37.67 –03 55 41.9	79.26	115.38	6.71	n.a.	Cet

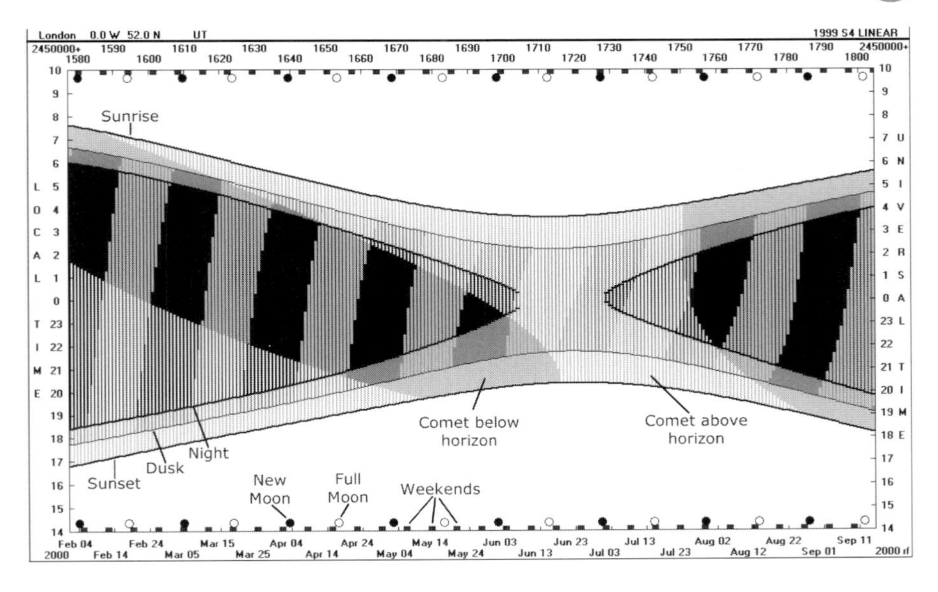

Figure 6.7 A screenshot of how the Grafdark program shows visibility circumstances. This plot is for Comet C/1999 S4 LINEAR in 2000. The plot shows when twilight or the phase of the Moon will interfere with observations. Courtesy Richard Fleet.

gets as it moves in towards the Sun. A lump of rock with no coma or tail would simply obey the inverse square law and K_1 would be set to five. Comets usually brighten rapidly as they approach the Sun since the coma increases in size as the nucleus' outgassing rate increases. In extreme cases K_1 can be as large as 20. Some example lightcurves are shown in Figure 6.8.

One word of caution, however. It is quite common for a comet to be given a large value of K_1 when it is first discovered implying that it will be very bright at perihelion. Depending on the age of the comet this initial high activity may not be sustained as the comet moves closer in towards the Sun. If so the resulting performance at perihelion will be disappointing. This can be particularly troublesome when the comet gets extensive early coverage in the media and then turns out to be a dud. This happened to the infamous Comet Kohoutek in 1973 and continues to happen occasionally when the media latch on to early speculation. People will never learn that comets are unpredictable things! One other thing to note is that the analysis of magnitude estimates often shows that a comet will have one law pre-perihelion and a different law post-perihelion. This reflects physical changes which occur in the comet as it passes near the Sun.

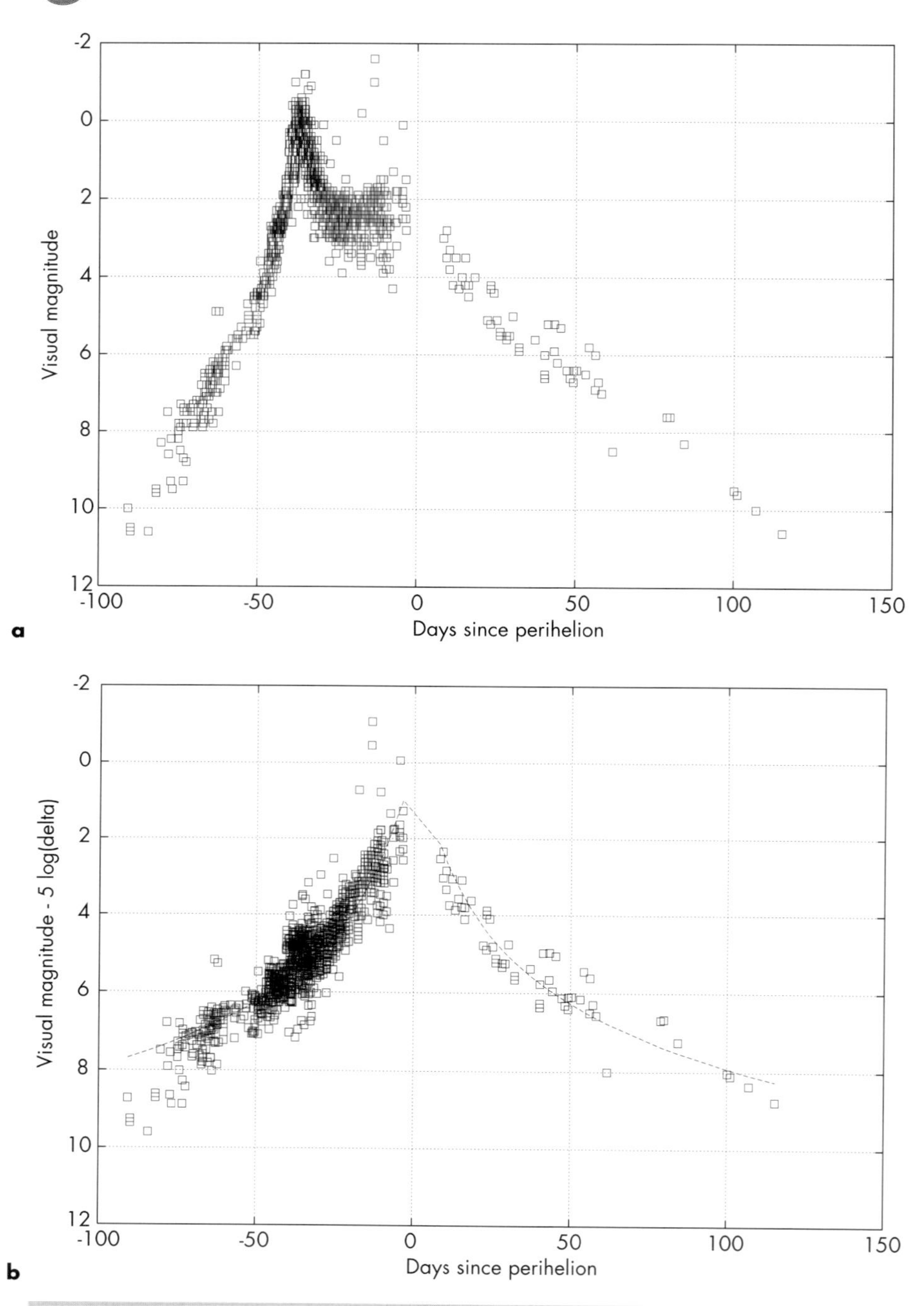

Figure 6.8 (a) A total of 1159 visual magnitude estimates of C/1996 B2 Hyakutake obtained by BAA/TA observers. The peak 35 days prior to perihelion corresponds to the comet's close approach to the Earth. In (b) the magnitude estimates have been corrected for the varying distance of the comet by subtracting 5 $\log_{10}(\Delta)$. The resulting plot shows how the comet would have looked if it was observed from a constant distance of 1 AU. The dashed line shows a good fit to the magnitude using $H_1 = 7.4$ and $K_1 = 7.5$.

Submitting Your Observations

You have seen how useful computers can be in characterising the motions of comets and in planning your observations. Until about ten years ago that was their main use in cometary astronomy but these days computers are also valuable tools for processing your observations.

We will discuss the details of computer based image processing in Chapter 8 but computers can also help with formatting your observations for submission to national bodies such as the BAA or the International Comet Quarterly (ICQ). These organisations have to deal with huge volumes of information and your observations will be processed much more quickly if you submit them in a standardised electronic format. Most of the time observations on paper are still welcome but if you can submit electronically you will be saving the coordinators a great deal of time and effort and you will ensure that your observations are entered into the databases without any transcription errors.

For visual comet observations there are a number of standard formats in use but the most common is the one adopted by the ICQ. This format is used in a modified form by the some other organisations such as the BAA and and an example submission is shown in Table 6.3. A full description of this reporting format is included on the CD-ROM but I suggest that you look at the web pages maintained by the various organisations for the latest information.

Comet astrometry is processed by the Minor Planet Center (MPC) and electronic submission is a necessity given the huge amount of observations that they receive. An example of the data record is shown in Table 6.4. Again a full description of the encoding is included on the CD-ROM. In all cases where you submit observations electronically it is important that you set your e-mail program to send text-only messages. Certain additions, such as HTML, which are selected by default in Microsoft Outlook Express should be switched off or your observations are likely to be rejected.

Images of comets should always be sent to the recognised astronomical organisations but there are

Table 6.3 Example ICQ format data for C/1999 S4. The example observation was made on 2000 July 22.9 by observer MIK. The eM/mm.m columns define the magnitude method (!V is photoelectric V band corrected for atmospheric extintion) and the magnitude estimate was 6.4 using the Yale Bright Star Catalog (YF) as the reference source. The observation was made with a 15.0 cm diameter, f/4 SCT (T). In this case the column xxxx indicates the CCD exposure. For visual observations it represents the power or magnification. The colums /dd.dd are the coma diameter estimate in arc minutes. In this case 8 arc minutes. The DC was 7. The colums /t.ttmANG are the tail length and position angle respectively (not reported in this case).

The full ICQ format is complex and many astronomical organizations use a simplified version. A full description of the coding can be found at http://cfa-www.harvard.edu/icq/ICQFormat.html

IIIYYYYMnL	YYYY MM DD.DD	eM/	mm.m:r	AAA.ATF/xxxx	/dd.ddnDC	/t.ttmANG	ICQ XX*OBSxx
1999S4	2000 06 19.06	!V	9.0 YF	15.0T 4a 60	+ 4.0 7	& 7 m270	ICQ115 MIK
1999S4	2000 06 29.13	!C	8.8 HI	40 D 3a 10	> 2.6 D7	> 8.2m270	ICQ115 ROD01
1999S4	2000 07 22.90	!V	6.4 YF	15.0T 4a060	+ 8.0 7		ICQ115 MIK
1999S4	2000 07 24.87			40 D 2a240	> 9 D 7	>27.0m 60	ICQ115 ROD01
1999S4	2000 07 10.06	S	7.9 TI	20.0T10	3.5 4	0.22 285	ICQ117 CRE02

Table 6.4. MPC format. This shows the standard format for submission of astrometry to the minor planet center. This was astrometry of C/2001 W2 Batters made by Denis Buczynski at IAU station 978 (Conder Brow). The astrometry lines consist of a packed designation (CK01W020) for the object, then the date and the RA and Dec. A full description of this record format can be found at the CBAT website http://cfa-www.harvard.edu/iau/info/OpticalObs.html

```
COD  978
CON  [ndj@blueyonder.co.uk]
OBS  D. Buczynski
MEA  D. Buczynski
TEL  0.33-m f/3.3 Newt.
NET  GSC (24 stars)
ACK  C/2001 W2 astrometry (correction)
     CK01W020   C2001 11 22.76704 18 15 22.15 +23 32 21.6          978
     CK01W020   C2001 11 22.77420 18 15 23.60 +23 31 51.0          978
```

now various websites that will archive and publish your images (some examples are included in the Appendix). If you are going to send an image as an e-mail attachment it helps if you adopt a standard file naming convention which includes the name of the comet, the observation date and your name. One format used by the BAA and TA in the UK is: **comet_yyyymmdd_observer.jpg** where comet is the name of the comet (e.g. 2000wm1), yyyymmdd is the date of the image, observer is a unique observer code (say your initials). If you have more than one image on a particular day just add "a", "b" etc to the end of the date field. An example would be 2000wm1_20011110_ndj.jpg for an image of C/2000 WM1 taken on 2001 November 10 by Nick James. It is a good idea to include details such as the telescope, scale, orientation, camera and wavelength as annotation on the image and it also helps greatly if images have a consistent orientation, preferably north at the top, east to the left.

Using Your Computer to Discover Comets

Finally in this chapter, one of the most extraordinary recent uses of computers by amateur astronomers has been the processing of images from the SOHO satellite with the aim of discovering comets. SOHO is a joint NASA/ESA spacecraft which was launched in 1995. It sits in space about 1.5 million km from the Earth and it produces what have become known as "the SOHO blue movies". These are timelapse movies (colour coded blue) taken by a coronagraph on board the spacecraft which show dynamic events in the Sun's outer atmosphere or corona. The field of view of the coronagraph is large enough that it frequently picks up comets passing close to the Sun (Figure 6.9). Professional astronomers had discovered many comets in this data but time and funding meant that the images were not checked very thoroughly.

In January 2000 Jonathan Shanklin, Director of the BAA Comet Section, was looking at one of the blue movies on the SOHO website when he spotted a Kreutz Group comet heading for the Sun. He reported it but assumed that it had already been noticed. In fact it hadn't been and so he was credited with the discovery. Since then a number of amateurs have written special software to process the SOHO images and the most successful hunter, Michael Oates, has discovered no less than 130 comets in the archival data! This is something that you can try for yourself since all of the

Figure 6.9 This image taken by the LASCO C2 coronagraph on-board SOHO shows Comet 96P/Machholz as it passed through the field of view on 2002 January 8. The Sun is blocked out by a central mask and Venus is visible to the lower right. At the time of this image the comet was only 22 million km from the Sun. Courtesy of SOHO/LASCO consortium. SOHO is a project of international cooperation between ESA and NASA.

data is freely available on the SOHO website. I should warn you, though, that these comets are named not after their human discoverer but after the spacecraft and so your discovery won't make your name an immortal part of the heavens.

So, computers can help you observe comets in many ways but we have spent long enough at the keyboard. It's time to return to the outdoors and to start looking even more closely at these fascinating objects.

Chapter 7

Comets In Close-up

The average comet is a small fuzzy object only a few arc minutes in diameter which requires a combination of large aperture and long focal length to show it well. In fact even Great Comets benefit from this "close-up" imaging. While Hale–Bopp showed spectacular tail features in wide-field images the stunning shells and jets surrounding its nucleus were only visible in images taken at large image scales using large telescopes.

To get good close-up images of comets you need three things: good optics, a good mounting and a good detector. Since comets are extended objects you will need fast optics to image them well. In fact, the ideal comet "camera" is a fast, long focal length, wide-field instrument. Such a combination is very difficult, but not impossible, to achieve and there is a range of suitable telescope types available. It used to be that film was the detector of choice for comet photographers at all focal lengths but CCDs are now the dominant detector for everything except the widest field images.

If you are very fortunate and money is no object then there are a huge number of wonderful gadgets available which will make your life much easier. Don't lose heart though if you don't have access to the very best equipment since enthusiasm can more than make up for limited resources. Most comets can be imaged perfectly well with a humble Newtonian or Schmidt Cassegrain Telescope (SCT) and it is far better to use the telescope and camera that you have than nothing at all. As with many things you should make the best of the circumstances that you find yourself in. I manage to obtain images of faint comets from a small back garden in a large, light-polluted town using a relatively low-tech Newtonian telescope and if I can achieve good results under these circumstances then so can you!

So let's start by looking at suitable optics and the important matter of field of view and resolution.

Image Resolution and Field of View

If you want to take high resolution images of a comet you need a large focal length (in excess of 500 mm) but then the field of view of your instrument becomes very small if you are using a conventional CCD. For example, my telescope has a focal length of 1.6 m and so it has a field of view of around $1^\circ.3 \times 0^\circ.8$ on 35 mm film but only $0^\circ.25 \times 0^\circ.17$ on my KAF-0401 CCD (a 6.9×4.6 mm chip). This chip has pixels which are 9 μm square and so each pixel corresponds to just over 1 arcsecond at the focal plane of my telescope. The seeing at my location is normally around 3 arcseconds and so the pixel size is considerably smaller than the average star (Figure 7.1(a). The pixel size of my system is probably a little too small for my circumstances although high resolution is good if you are planning to do astrometry on the image. The image is said to be *oversampled*. Quite often I will use my chip in 2×2 binned mode so that the pixels are effectively 18 μm square. This has the advantage of increasing the sensitivity of the chip since the pixels have four times the area but the total size of the CCD and, of course, its field of view remain the same.

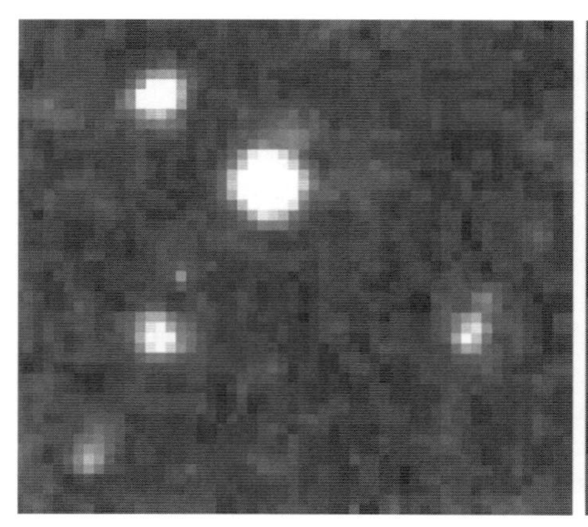

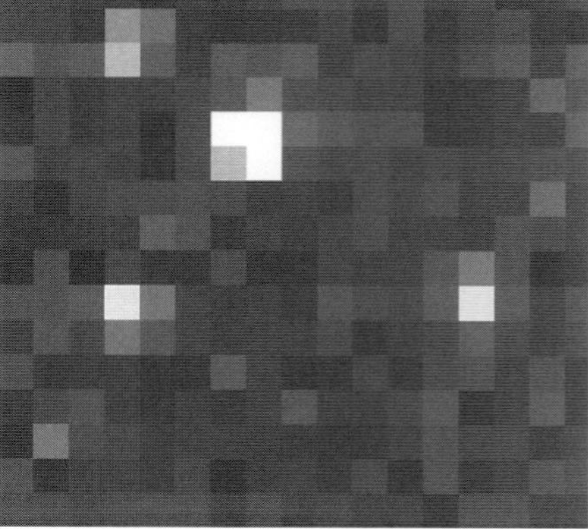

Figure 7.1. These two images show the effect of pixel size on star images. In the left hand image the stars are oversampled, in the right hand image they are undersampled.

My CCD is fine for the vast majority of comets but once or twice a year a more spectacular comet will come along which won't fit in the field of view of the chip. I could use a larger CCD but the price of suitable cameras escalates rapidly as the imaging area increases. An alternative is to reduce the focal length of the optical system but if I go too far the image will be *under-sampled*. If I used my CCD on a telescope with a focal length of 0.4 m the field would be four times larger but each pixel would represent almost 5 arcseconds at the focal plane. Stars would now be smaller than the pixels they would start to look "blocky" (Figure 7.1(b)). If you are planning to do astrometry with your images and you have the freedom to select equipment I suggest you aim for a combination of optics and camera which gives you a resolution two to three times that of your average seeing disk. Alternatively if you intend to do photometry or image comets to detect large-scale structure a scale of up to 4 or 5 arcseconds per pixel would do.

Most of us will not have the freedom to change telescope or detector to get the optimum sampling. If this is the case you should make the best of what you have got and perhaps investigate modifying the focal length using a telecompressor. The Optec f/3.3 compressor for instance will reduce the focal length of an f/10 SCT by a factor of three and increase the field of view accordingly. The faster system will also be much more suitable for comet imaging although off-axis aberations using a focal reducer can limit the field of view on large chips. Alternatively you may already have a telescope of a particular focal length but don't yet have a CCD camera. If this is the case you should select a camera with pixels which match the size criteria listed above. Common CCD chips have pixels ranging from 7 μm to 24 μm so there is a good range to choose from.

Wide-Field Optics

Since comets are extended objects the image brightness of the coma or the tail is dependent on the f/ratio of the optics used. This is just the same as you are used to with everyday photography and different to what happens with stellar photography where the brightness of stars in the image is dependent only on the aperture rather than the f/ratio. A comet imaging telescope should be as fast as possible but it should also have a wide, distortion-free field.

If you want a wide field and high resolution you will need a large detector. Whether this is a CCD or film the optical instrument will need to illuminate this large focal plane area with minimal distortion and vignetting. For really large comets the only practical choice of large area detector at present is film. With the increasing size of CCD chips the use of film in even this application will decline but we did see a rebirth of film-based comet photography in 1996 and 1997 when the Great Comets Hyakutake and Hale–Bopp were at their best. In these cases even 35 mm film was not large enough and advanced observers were using medium format film (6 cm square) or even large film sheets (up to 8-inch × 10-inch) to get a combination of good resolution and a huge field of view.

You may not be able to get a wide field simply by placing a large format camera at the focus of your telescope. This is because many optical arrangements suffer from off-axis distortions which limit the usable field no matter how large a detector you use. As mentioned in Chapter 3 the common Newtonian telescope suffers from off-axis coma which severely restricts the usable field. The coma-free field of a Newtonian is proportional to the inverse square of the f/ratio. An approximation for the size of the comatic star image is given by:

$$C = \frac{50.x}{FR^2} \mu m$$

where x is the distance off-axis in the focal plane (measured in mm) and FR is the focal ratio. Compare the size of the comatic image with the pixel size of common CCDs and you will see that fast Newtonians are only suitable for small focal plane detectors. You will see that coma is already apparent using 35 mm film at f/5 and that at f/3 it would be a problem even with a relatively small CCD. If you want to do wide field work with a fast Newtonian there are some commercial coma correctors on the market such as the Teleview Paracorr which do a good job of correcting coma at focal ratios as fast as f/3. Also Meade have recently brought out a range of Schmidt-Newtonians which use a corrector plate to increase the size of the coma free field for a given f/ratio. The 10-inch, f/4 LXD55 looks to be a good choice for comet imagers and it is reasonably priced at under $1K in the US.

One of the most commonly used telescopes these days is the Schmidt–Cassegrain (SCT). These instruments offer excellent value but they are really too slow for wide-field comet imaging. The normal focal ratios of f/10 and f/6.3 can be reduced using a telecompressor but the field is then severely limited by off-axis aberrations. If you really want a wide field, fast telescope for that rare Great Comet you will need to buy a specialised instrument.

In the years after the Second World War a large number of ex-military aerial reconnaissance cameras became available to astronomers and these were ideal for comet photography using large format film. They had large focal lengths, were quite fast and illuminated huge focal plane areas (up to 8-inch × 10-inch). Such lenses are still available at reasonable prices on the second hand market but they are of variable quality and it is getting increasingly difficult and expensive to obtain the sheet film that is needed to use them. These cameras are also unwieldy and difficult to use. Similar cameras designed specifically for imaging the night sky were used by dedicated comet photographers to obtain spectacular images of comets from the 1950s through to the late 1980s (Figure 7.2). The modern equivalent of

Figure 7.2. This photograph of C/1956 R1 Arend-Roland was taken on 1957 April 30.9, It is a 67 minute exposure on a Kodak Oa-O plate using a 6-inch diameter Cooke triplet lens of 26-inch focal length (i.e. around f/4.3). Reginald Waterfield.

the large astronomical camera is the fast, short focus, apochromatic refractor. These telescopes have the advantage that they can be used both visually and as a very high quality photographic lens capable of illuminating medium format film. The results are superb (Figure 7.3) but the instruments are very expensive. The 130 mm, f/6 EDF apochromat from Astro-physics costs $4K for the tube alone.

Returning to reflective optics the hyperboloidal astrographs manufactured by Takahashi are particularly suitable for imaging large comets. These astrographs, which are sold under the name Epsilon, fully illuminate a 6cm square medium-format frame with negligible distortion. They are available in versions ranging from 180mm aperture f/3.3, to 300 mm aperture f/3.8, so the 300mm aperture would give you a field of view of over 3° on medium-format film. Even 35 mm film gives very good results and Martin Mobberley has used the 160mm aperture f/3.3 instrument to record Hale–Bopp in all its glory (Figure 7.4). The Epsilon has the advantage that the focal plane is flat, that is over the field of view all of the stars off-axis come to a focus in the same plane. This means that you can use a standard film back and roll film to obtain your images.

An alternative to the astrograph is the Schmidt camera. This is the ultimate in a fast, large format, large aperture instrument. The basic optical design was worked out by

Figure 7.3. Comet C/2001 A2 (LINEAR) on 2001 June 30 at 10^h 00^m UT taken with an Astro-Physics 130EDF, f/6.0 apochromatic refractor and Finger Lakes Instruments CCD camera. 33 frames of 1 minute each were combined to form this image. The camera was guided using an ST4 autoguider attached to a Celestron 80 mm guide scope. The field of view is approximately $1^\circ.8 \times 1^\circ.3$ with south up. Wil Milan.

Figure 7.4. Comet C/1995 O1 Hale–Bopp on 1997 March 30 between $20^h\ 46^m$ and $20^h\ 56^m$ UT. Takahashi E160 (160 mm aperture, f/3.3) using Kodak Ektar Pro Gold 400. Field of view $2^\circ \times 3^\circ$. Martin Mobberley.

Bernhard Schmidt in 1930 and it is still unsurpassed when it comes to implementing wide-field, fast optics. In fact the famous Palomar Sky Surveys were conducted using a 48-inch aperture Schmidt camera. While they can achieve spectacular results the optical arrangement does have the disadvantage that the focal "plane" (more properly the focal surface) is not flat. This means that the film must be held against a specially shaped focal plate so that it forms the correct shape to match the focal surface. Schmidt cameras are usually described by three numbers written as a/b/c where a is the aperture, b is the diameter of the primary mirror (this is always larger than the aperture) and c is the focal length.

Various manufacturers sell Schmidt cameras in the commercial market but the Meade LX200SC is probably the best example. It is a 30 cm aperture, f/2.2 camera which takes 6 cm square film and provides an unvignetted field of view of 4°.3. All of this for around $25K! Similar cameras have produced spectacular results on bright comets since the combination of large image scale, wide field and very fast optics is unbeatable (Figure 7.5). The curved focal plane means that film has to be cut into single frames for loading into the camera and so Schmidt cameras are definitely for the experts only. In any case most comets are far too small to be suitable targets and it

Figure 7.5. Comet C/1995 O1 Hale–Bopp obtained using a Schmidt camera. 1997 March 11, 04^h 05^m UT, Hypered Technical Pan film with a 20/22/30 cm Schmidt. Lennart Dahlmark.

is unlikely that one suitable comet a decade could justify such an expensive camera.

A variation on the theme is the Baker–Schmidt. This isn't quite as fast as a true Schmidt but it has the advantage of having a flat focal plane. This type of camera is widely used by active comet observers with Herman Mikuz in Slovenia producing excellent work with a 20 cm, f/2 version (Figure 7.6).

All of these specialised optical instruments can produce spectacular comet images but let me repeat: It is far better that you make use of the telescope that you have even if it is not entirely suitable. Practically every telescope can be used for comet imaging. The same, however, cannot be said about the mounting. This is the one area that you really do have to get right. In fact, if resources are limited, my view is that the mounting should always get priority over the optics and detector.

Telescope Mounts and Guiding

I speak from experience when I say that you will need a good telescope mount or your attempts at comet

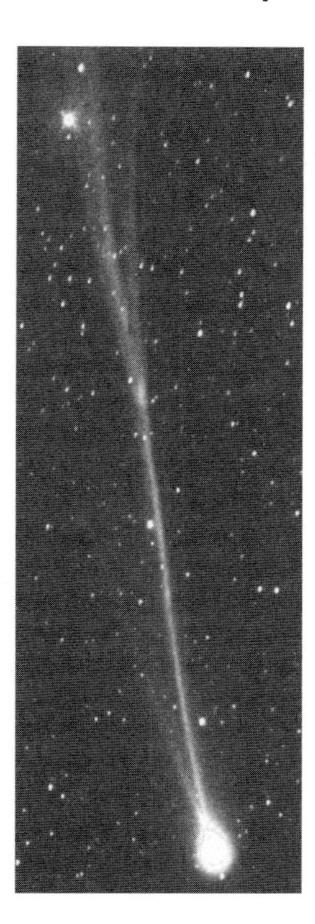

Figure 7.6. Comet 122P/de Vico on 1995 October 1 at $03^h\ 18^m$ UT. A mosaic of two 180s exposures taken using an ST-6 at the focus of a 20 cm, f/2 Baker–Schmidt. Herman Mikuz, Crni Vrh Observatory, Slovenia.

imaging will become a nightmare. For good images the mounting and drive must follow the comet with an accuracy of 2–3 arcseconds for the duration of your exposure. This is a tall order but the very best telescope mounts, such as the Paramount GT-1100ME from Software Bisque claim to allow unguided exposures of up to four minutes with focal lengths of 2.5 m (Figure 7.7). The Paramount is an example of a very accurate *robotic* telescope mount and it can even be programmed to follow the motion of a comet against the background stars. They are expensive (around $8.5K at the end of 2001) but worth every penny if you can afford them since they greatly simplify the twin problems of finding and guiding on a comet.

More conventional mountings, such as the ubiquitous Meade LX200 mount are not so accurate. Depending on your mount, periodic error in the drive system and other imperfections may limit your exposures to a minute or less. In addition most comets are moving quite rapidly with respect to the background stars and so in some cases your exposure may be limited by the motion of the comet itself. With CCD cameras short exposures of 30 seconds to a minute are often more than adequate for astrometric work. However, you may want to use longer exposures if you are after images which will show coma and tail structure or if you want to perform photometry. In this case you have two alternatives.

The simple approach and the one which is usually adopted by many CCD imagers is to take multiple short exposures of the comet and then align and stack them in the computer after your observing session. You should choose an exposure length which gives you a high confidence that your uncorrected drive will be adequate and that the comet's motion is small. This approach also has an advantage in the case of very bright comets in that you can avoid saturating the brightest part of the coma by using shorter exposures.

There is a disadvantage to this approach in that each of the short images is subject to the read-out noise of the CCD electronics whereas a single long exposure has only one unit of read-out noise to contend with. Assuming that the exposures are relatively short so that they are limited by read-out noise the signal-to-noise ratio (SNR) of the final stacked image only improves by the square root of the number of images that you have taken. This means that the equivalent exposure of n frames of exposure t seconds is only $\sqrt{n} \times t$ rather

Figure 7.7. A Paramount robotic German equatorial mounting. This type of telescope mounting is ideal for comet imagers since the mount can be controlled by a computer and the drive is extremely accurate.

than $n \times t$. To put it another way, to get the same SNR as a single 100s exposure you would need to take 25 exposures of 20 s each or 100 exposures of 10 s each. While this can usually be tolerated since it removes the need for guiding it does mean that very long equivalent exposures become impractical.

The alternative is to offset guide. Let me say now that offset guiding is probably one of the most difficult aspects of prime-focus cometary imaging which is why most CCD observers prefer to use the stacking approach despite the effective exposure loss. Of course, while offset guiding is optional for CCD detectors it is mandatory for photography since stacking multiple film exposures is generally out of the question.

The general idea of guiding is to "close the loop" so that the telescope actively follows a specified guide star. A guiding image is monitored by a human observer or electronic autoguider and telescope drive rates are controlled to keep the guide star at the same point in the image. There are generally two arrangements for obtaining the guiding image. Traditionally, guiding has been performed using a separate guide telescope bolted onto the side of the main instrument. Such guiding arrangements are still in use today and are essential if the main instrument is an "imaging only" system such as a Schmidt camera. This arrangement works well if the telescope and its optics are rigid so that there is no flexure between the two instruments. The mounting of the guide telescope will allow it to be adjusted so that you can search out a reasonably bright guide star some way from the target comet.

Guide telescopes are not usually suitable for SCTs since these telescopes often suffer from image shift caused by their optics moving within the tube assemblies ("mirror flop"). An alternative which overcomes this problem is to use a small mirror or prism to pick off some light from the field of the main instrument and feed this to a guiding port. The mirror is sufficiently far off axis that it doesn't affect the imaging system and these units are called off-axis guiders (OAGs). The mirror can be moved around the field to find a guide star but the choice of star is usually more limited than you would have with a separate guide telescope. An OAG does however have the advantage that there can be no flexure between the guiding port and the focal plane image (assuming a rigid pick-off mirror) and so OAGs are now much more common than separate guidescopes.

To guide manually you place a relatively high powered illuminated reticle eyepiece (say a magnification of twice the focal length of the imaging instrument measured in cm) into the guide port and then press the RA and Dec slow motion controls on the handset to maintain the star at a selected point on the reticle. CCD autoguiders, such as the SBIG ST-4, have taken away the hassle of visual guiding for those lucky enough to have one. A small CCD chip is used to image the guider field using short exposures and the selected guide star is monitored electronically from one frame to the next. If the star moves away from its nominal position the computer issues commands to move the telescope to bring the star back to the centre just as a human observer would if they saw the star moving off the guide wire.

Some CCD cameras, such as the ST-7, have a separate guide chip built into the camera so that there is no need for an OAG at all. The disadvantage of this approach is that the secondary guide chip is in a fixed position relative to the imaging chip so the choice of guide stars is limited. An alternative approach has been adopted by the Starlight Xpress range of cameras. Their cameras use an interline CCD and so it is possible to read out alternate lines of the image for use in guiding without affecting the other lines which are still integrating the image. Their STAR (Simultaneous Track and Record) software works very well at the expense of a reduction in sensitivity of the CCD although this can be counteracted by doubling the exposure (Figure 7.8).

All of this is fine if the comet is bright and has a well defined false nucleus since it is possible to guide directly on that as if it were a guide star. In many cases the comet will be too faint or diffuse for this method and you will need to "offset guide" on a real star in a way that takes account of the comet's motion. To start with you will need to know the speed and direction of the comet's motion relative to the background stars. This is normally given in arcseconds per hour in the direction of a certain position angle and can be obtained from the ephemeris (see Chapter 6).

Traditionally offset guiding was performed using a micrometer eyepiece. The guide wire was moved slowly at the correct rate in the opposite direction to the comet's motion and the guide star was maintained on the guide wire. More recently, with the availability of illuminated reticle eyepieces (such as the one described in Chapter 4) it is possible to align the reticle along the direction of the comet's motion and then offset the guide star by one division in a set interval of time.

Figure 7.8. Comet C/2000 WM1 LINEAR taken using an MX7C colour camera with the STAR2000 autoguiding system following the comet's false nucleus. The camera was attached to a C11 at f/4 and the exposure was 2 minutes.
Terry Platt.

The time taken to offset by each division is determined by the comet's rate of motion. You will also need to know the actual distance between the small divisions on your reticle in arcseconds on the sky. This can be determined using the technique described in Chapter 4. You now calculate the time to move one division along the reticle from:

$$t = \frac{3600 \times d}{m} \text{sec}$$

where d is the reticle division in arcsec and m is the comet's motion in arcsec/hr taken from the ephemeris. During the exposure you will need to move the guide star along the reticle at this rate in a direction aligned 180° from the PA of the comet's motion. The easiest way to do this is to make a tape recording which lasts for the length of the exposure and on which you record a commentary describing where the comet should be on the reticle every thirty seconds or so.

Take it from me, these methods of offset guiding are really hard work and I was pleased to see the back of them when I moved from film to CCD. In reality there probably isn't any need for them now since any comet big enough to be worthy of film will usually be bright enough that you can guide directly on the nucleus.

Since offset guiding was one of the most unpleasant aspects of cometary photography people came up with some particularly clever inventions to avoid it. One of the best is the focal plane platen guider. In this gadget the observer guides on a fixed guide star but the detector is moved slowly across the focal plane to counteract the comet's motion. Computer controlled platen guiders driven by stepper motors work very well and an example is shown in Figure 7.9. There are a

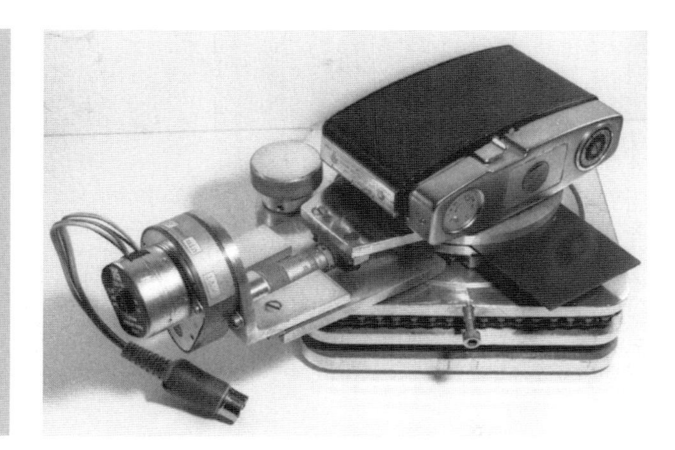

Figure 7.9. This focal plane positioner was designed by Ron Arbour in the days of film photography but it is just as useful today. A computer controls the motion of film or CCD in the focal plane in order to counteract the comet's motion. The guiding system can then track on a fixed guidestar.
Ron Arbour.

number of commercial companies which sell motorised micropositioners which you can adapt to do this sort of thing (see the Appendix). If you use this arrangement it is possible to use a commercial autoguider since it will be guiding on a fixed guide star while the detector moves. While you might think that it would be relatively easy to write software for a CCD autoguider to enable offset guiding none of the commercial guiders seem to be able to do this.

Astronomical CCD Imagers

The third requirement of the comet imaging system after good optics and a good mount is a good detector. In Chapter 5 we encountered CCD cameras. You will remember that non-astronomical CCD cameras are not cooled and they are not usually capable of taking the long exposures that we need. If you want to do quality comet imaging with a CCD camera you will need to have a specialised astro camera. This will be cooled to allow exposures of many tens of minutes but it will come at a significant increase in price over conventional digital cameras.

If money is no object you should select a camera which has a large field of view, a pixel size which is matched to your optical system, a very high sensitivity and has at least 16-bit read-out resolution. Sadly, for most of us, the choice of camera will also be strongly influenced by how much it costs! This is not the place for an extensive review of the cameras available on the market and if you are interested in buying such a camera I suggest that you look at some of the books listed in the Appendix or read reviews in the astronomy press. In the limited space that I have here I restrict my discussion to particular points which are important to comet imaging.

The *quantum efficiency* (QE) or sensitivity of the camera is a key factor in determining how faint you can go with a given setup. There are a number of features of CCD design which tend to reduce the QE. In particular the metal electrode structure that is used to form the pixel wells will shield the sensitive regions from photons and so sensitivity is lost. In addition the photons need to be able to burrow their way far enough into the chip to be detected. It is possible to take conventional "front-illuminated" CCDs and very carefully etch away the substrate behind the pixel wells to leave a very thin chip which can be illuminated from the back. In this kind of CCD the photons enter through the back surface of the device and reach the light-sensitive regions after travelling through the substrate. Quantum efficiencies of this kind of CCD can be very high (above 90%) and the sensitivity to shorter (blue) wavelengths is dramatically improved. Quite often the back surface will be treated with an anti-reflection coating to improve the QE still further.

Back-illuminated CCDs are much more expensive than their front-illuminated cousins since the etching process is rather difficult but they do offer stunning performance. The best known commercial example is the Apogee AP-7 which is based on a thinned s 512 × 512 SITe chip with 24 μm pixels. This camera is capable of reaching 19th magnitude in 30 seconds on a 14-inch telescope and costs around $5.5K at the end of 2001. Apogee also offer a camera based on a thinned 1K × 1K Marconi chip with 13 μm pixels. The AP-47p costs around $7.5K. In comparison a good quality front-illuminated camera such as the SBIG ST-7E costs around $2.6K. This camera is based on the Kodak KAF-0401E chip with 768×512 9 μm pixels. A double sized version, the ST-8E, is based on the KAF-1602E with 1536 × 1024 pixels of 9 μm so it offers twice the field of view but costs more than twice the ST-7E.

Another crucial factor is the *full-well capacity* of the chip. The best comets have a huge variation in brightness from the centre of the coma to the extremes of the tail. If we are to record all of these features we need to expose so as not to saturate the brightest parts of the image but still record the faintest. The full-well capacity is a measure of how many electrons can be stored in a pixel before it overflows and so it is a good indication of the difference between the faintest and brightest parts of the image that can be faithfully recorded (the dynamic range). The full-well capacity is usually better for cameras with larger pixels and a good camera should have something like 100,000 electrons or more. The number of bits used to store the digital image is also important in this regard and practically all modern cameras have 15 or 16-bit resolution.

If you plan to do cometary photometry the pixels in your CCD should have a linear response to incident light, that is twice the intensity should lead to twice the signal. If parts of the field of view are saturated the pixels will exceed their full-well capacity and the excess electrons will have to go somewhere. As I mentioned in Chapter 5 some cameras have anti-blooming drains to remove these excess electrons but these should be avoided in scientific cameras since they reduce the linearity, sensitivity and full-well capacity of the detector.

As an alternative to buying a camera you could always build your own. The Audine cameras (based on the same KAF-0401E/1602E chips that are used in the ST-7 and ST-8) and the Cookbook camera (based on the Texas TC-245) are two examples of cameras which can be built at home using instructions which are publicly available (see the Appendix). There are even some companies which manufacture and sell these designs commercially. For example a KAF-0401E based Audine camera is available in the UK from Rockingham Instruments at around £900.

Other Equipment

Comets are colourful objects and much information can be obtained by imaging them in particular colour bands. You may also want to obtain full colour images if the comet is bright. While some astronomical CCD cameras offer a "one-shot" colour mode they are not suitable for scientific work and the usual technique is to take three images of the comet through red, green and blue filters and combine these later at the image processing stage. All of this is greatly simplified if you use a filter wheel between the camera and the telescope and many of these allow you to select a particular filter under computer control. It is most important that the filter positioning is very accurate so that flat fields taken with a particular filter can be used to correct for dust and other imperfections throughout the night. A good example of a motorised filter wheel is the "Custom Wheel" from True-Technology Ltd. This costs around £600 in the UK.

Another very important item of equipment is often neglected. If you use the standard focuser available on most telescopes you will find that it is almost impossible to get a sharp focus. I recommend the use of a zero image shift focuser such as the NGF series from JMI. These are quite expensive (around $300) but they remove much of the hassle from focusing. Motor driven focusing is a great luxury and it can remove much frustration especially if you can focus the telescope remotely from your computer.

CCD Calibration Frames

Comets are usually very faint, low-contrast and diffuse objects so the coma and tail gradually merge with the background sky. Even a bright comet such as Hale–Bopp had faint extensions to its tails. CCD images can be processed to bring out the very faintest detail but to do this we must be able to correct for the various imperfections in the camera and optical system. The first step, before we even find the comet, is to take the *calibration frames* which will allow you to make these corrections as described in the next chapter.

Three types of calibration frame are required. A *flat field* is a short exposure of an evenly illuminated field (often the bright twilight sky) taken with the camera and optical arrangement that will be used to image the comet. If the camera and telescope were perfect, every pixel would have the same brightness value. In fact a variation of sensitivity from pixel to pixel means that the flat field contains a range of brightness values. In addition the telescope will probably vignette the field and so the amount of light falling on the chip will fall off towards the edges. Finally, unless the camera is perfectly clean, there will be specks of dust which will cast shadows on the chip. The flat field records the variations in pixel sensitivity due to these effects. Flat fields are particularly important in comet imaging since the extremities of a comet's tail may be only a only a few percent brighter than the sky background. This is especially true if you live in an area where the light pollution is significant. If you are using filters you will need a separate flat field for each filter used.

The flat field is the most difficult calibration frame to get right. The conventional way of obtaining it is to take a short exposure of a uniformly illuminated patch of sky at twilight. You should obtain the image using the normal optical arrangement used for comet imaging and, for the best results, the camera should not be detached from the telescope between taking the flat field and taking the comet image. The sky should be bright enough so that you can keep the exposure short (less than a second) but not so bright that any part of the image is saturated. To reduce the various sources of noise in the flat field it is best to take 10 or 20 exposures of the same length that can be processed into a single frame later. This can make it rather difficult to obtain a good flat field since there is only a ten or twenty minute window during evening or morning twilight when the exposures can be made. This is especially the case if the weather is unpredictable since you may have a sudden clearing which allows you to image a comet late at night even though you had no chance to make a flat field earlier on. Another problem is that you will often find, even though the sky appears to be quite bright, that there are some stars in your flat field. These can be removed using computer processing as long as there are not too many of them.

There are a number of alternative ways of obtaining a flat field which do not require you to be ready during that crucial twilight window. Basically you need to arrange for a perfectly uniform illumination across the aperture of your telescope. If you have an observatory you may find that you can obtain a good flat field by imaging the illuminated inside of the dome. Alternatively you can make a diffusing light box such as the one shown in Figure 7.10. This fits over the front of the telescope and it allows you to obtain a flat field at any time. The difficulty is ensuring that the illumination really is uniform to the 1% level. Most people achieve this by ensuring that the light is scattered many times within the box before being presented to the telescope.

The next type of frame is much easier to obtain and it can be taken just after the flat field. This is the *bias frame*. The data from most cameras is offset from zero by some fixed amount which is dependent on the readout electronics. The bias frame is a zero length exposure frame obtained with the shutter closed or telescope capped. If your camera software can't do zero length exposures select the shortest exposure that you can. As with the flat field take 10 or 20 bias frames for later averaging to reduce the effect of noise.

Finally the *dark frame* is normally taken during the night near to the time that the actual comet images are obtained. This is an exposure made using the same exposure time as the comet image but with the shutter closed or the telescope capped. Since no light is allowed to fall on the CCD each pixel contains only

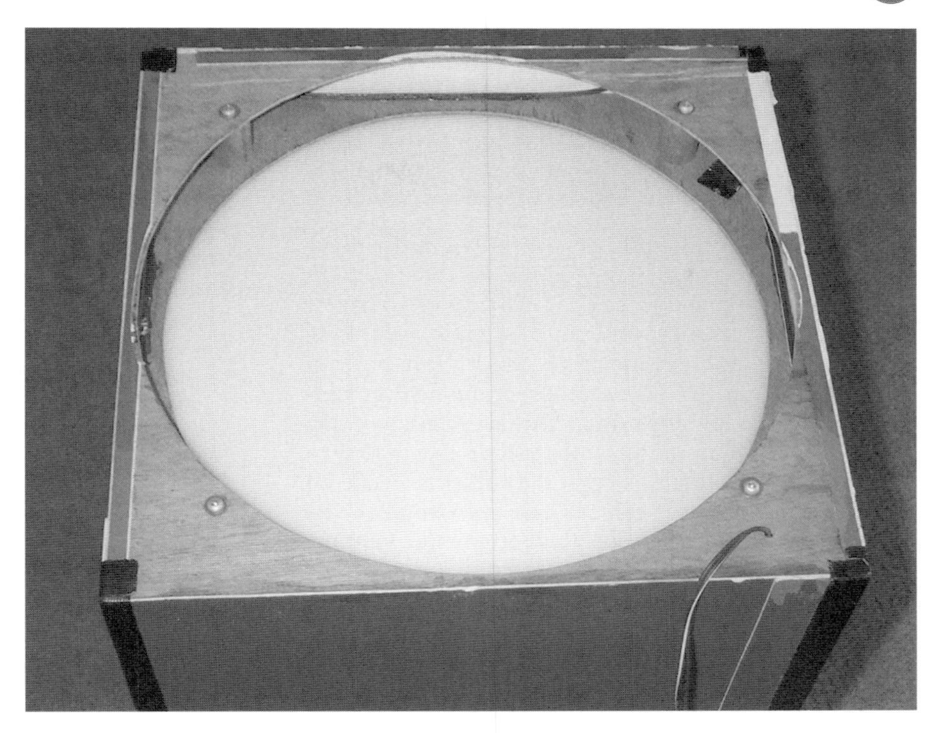

Figure 7.10. A diffusing light box can be used to generate flat fields. The illumination must be uniform over the telescope's aperture for this to work. This light box uses diffusing perspex and four 2W miniature tungsten bulbs fed from a 12V power source via 15W, 10Ω resistors. The box is held in place on the telescope tube by four corner dowels (right). Steve Goldsmith.

the thermal noise associated with the exposure. It is best if the camera temperature does not vary too much between the image frames and the darks but this is not absolutely necessary as I will explain in the next chapter. As with the other calibration frames, if time permits take 10 or 20 darks for subsequent processing.

Finding Faint Cometary Objects in Narrow Fields

After taking the calibration exposures you must move your telescope accurately enough to place the comet's image on the chip and since CCD cameras often have a very small field of view this can be a quite frustrating exercise. Your task is made much easier if you have an accurately set-up GO-TO telescope. If the field of view of your system is larger than the expected pointing error of your mount the telescope will be able to slew to the specified coordinates so that the comet is visible on your monitor. If this is not the case then you can use the technique described below to fine-tune the pointing.

If you don't have a GO-TO facility on your telescope you will need to use rather more primitive means to find the comet. Some older telescopes come equipped with setting circles and these can be a very useful aid in the absence of anything better. With setting circles I recommend that you offset the telescope from a nearby bright star rather than dial in absolute coordinates. Using this method you take the comet's coordinates and then select a nearby bright star (say up to 30° away). Calculate the offsets from the star to the comet in RA and Dec and then line your telescope up on the star. Finally use the setting circles to move the appropriate offset to get to the comet.

Most setting circles and some GO-TO mounts will not be accurate enough to get the comet in the field of view first time and I suggest that you prepare a chart with a field size which is consistent with your expected pointing error (Figure 7.11). You should be able to recognise star patterns in the CCD image even if the telescope isn't pointing in exactly the right direction and you can then offset to the actual position of the comet. This is made much easier if you set your camera so that you always have a consistent image orientation (e.g. north at the top). Don't forget that some comets move rapidly with respect to the background stars and so the comet's position should be marked on the chart for the expected time of observation.

Once your telescope is pointing in the right direction the comet should, in most cases, be visible as a fuzzy object on the monitor although some really faint

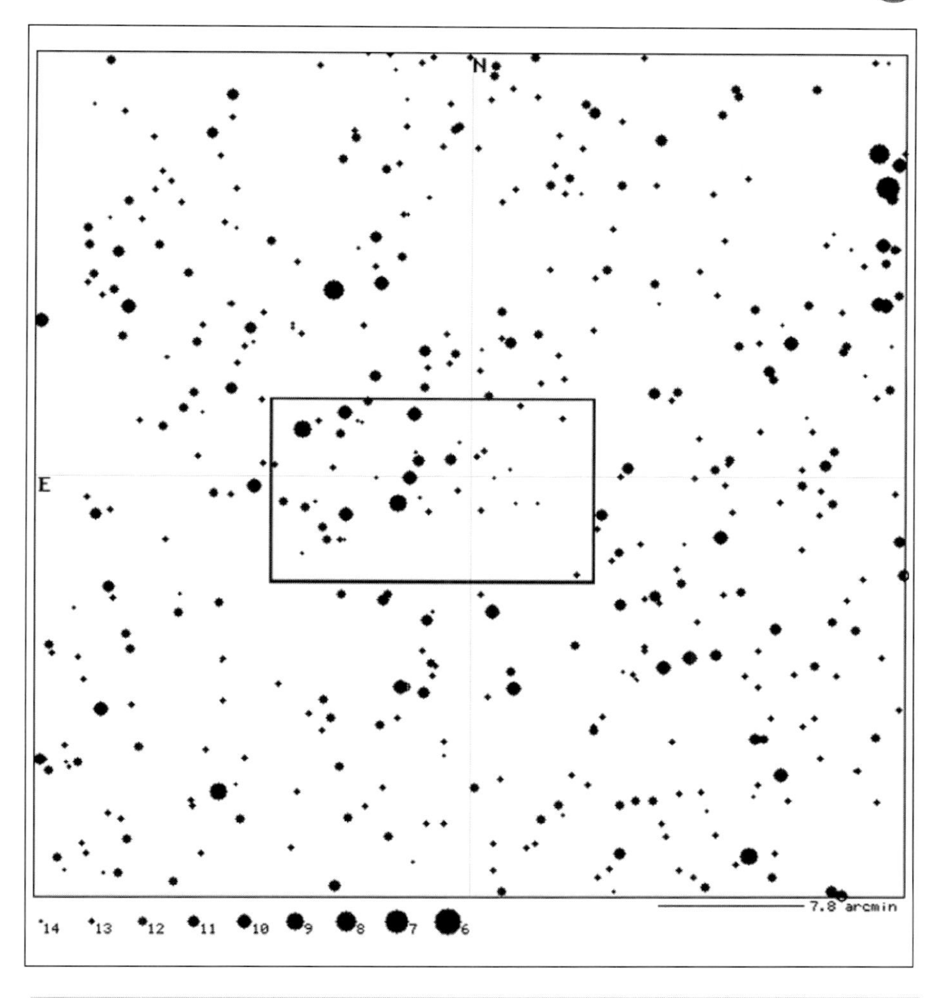

Figure 7.11. a A finder chart for C/2001 A2 LINEAR on 2001 September 10.85 based on the Guide Star Catalog. The field of view of the CCD is marked. **b** The actual CCD image showing the comet at around magnitude 16.
(Figure 7.11b is on the following page)

comets are hard to find in the short exposures used for lining up the telescope. It is worth practising all of this on deep sky objects before trying to find fainter comets. Deep Sky objects don't move and suitable objects are available all of the year around. Simply pick a galaxy which has a catalogued magnitude of 13–14 and then try searching it out with your telescope and camera.

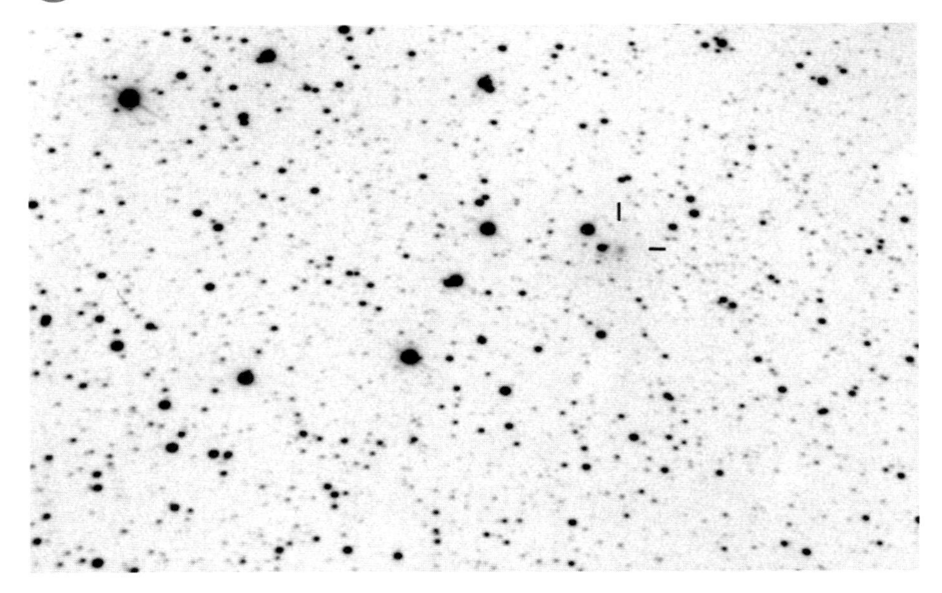

Figure 7.11b.

Obtaining the Image

Before you take the image you must ensure that the camera is properly focused. Most CCD control packages allow you to select a particular star in the image. A small area around this star can then be read out and displayed very rapidly as you adjust the focus to maximise the signal. All of this is greatly simplified if you have a motor-driven focuser which can be controlled from the same computer that you use for imaging. Some software packages can then perform the focusing automatically by rapidly reading out a small area of the image around a selected star and then adjusting the focus to maximise the peak of the star's brightness profile.

You are now ready to begin imaging the comet. If your objective is to perform astrometry on the images you should select an exposure which is as short as possible consistent with getting a strong image of the centre of the coma. In all other cases I would suggest that you select an exposure which, though longer, does not saturate the coma since there is often a wealth of detail near to the false nucleus which is lost if the exposure is too long. You can use the histogram

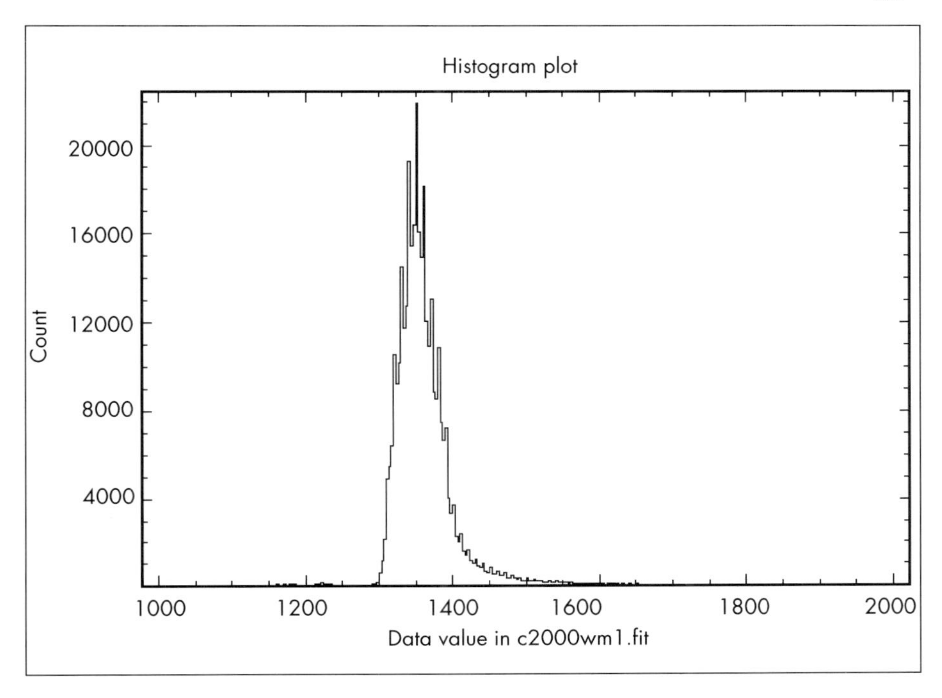

Figure 7.12. This histogram shows the distribution of brightness values recorded by the CCD. The horizontal axis is the brightness value and the vertical axis shows the number of pixels having the given brightness value. In a 16-bit camera the maximum brightness value is 65536. If any areas of the image are saturated the histogram would show a second peak at a brightness value of 65536. This particular histogram has a horizontal axis which is scaled between the minium and maximum pixel values in the image. It shows that there are no saturated pixels in this image.

function of your camera control software to check that your selected exposure is correct (Figure 7.12). Bear in mind also that, if you are not offset guiding, the exposure should be short enough that the comet is not appreciably blurred by its motion relative to the star background. Don't forget that you can take multiple frames, one after the other, and combine them later.

The raw comet images that you obtain during the night will probably look rather bland but don't despair. They probably contain some very interesting details but these won't be visible until you complete the image calibration and processing stages and that is where we go next.

Chapter 8

Improving Your Image

In the days before digital imaging it was very difficult to process comet photographs to bring out all of the information that they contained. You could choose a particular developer, film and print paper to achieve the desired contrast and you could selectively adjust the density of the print at different points in the image by moving a sheet of cardboard under the enlarger but that was about it. It was particularly difficult to print a photograph of a comet to show faint tail detail without burning out the bright coma and the stacking of multiple images required highly specialised techniques. All of this changed when images became available in digital form.

The Digital Darkroom

Digital image processing can be used to extract the maximum amount of information from your hard won comet images. The "digital darkroom" inside your PC is far more powerful than the old fashioned photographic darkroom and much more fun to use. While you can't extract more information from an image than what is there in the first place you can certainly process the image so that the information is easier to comprehend. Take, for example, the raw image of Hale–Bopp shown in Figure 8.1(a). This shows a fairly bland, low contrast object. We could simply stretch the contrast but this only succeeds in saturating all of the interesting details. A commonly used technique is to use an *unsharp mask* to bring out small-scale detail. The mask (Figure 8.1(b)) is simply a blurred version of the original. If we divide the original image by the mask and then stretch the contrast we get the result shown in Figure 8.1(c). Finally we apply a filter to the image to remove some of the noise (Figure 8.1(d)). All of these techniques are easy to perform in your PC given the right techniques and software.

CCD images are already in the correct format for image processing but film images need to be converted before they can be used. You may already have a scanner which can be used to scan photographic prints into a digital format. While this is better than nothing it is not usually satisfactory for comet images since the

a

b

Figure 8.1. Image processing of the central region of C/1995 O1 Hale–Bopp. **a** Contrast streched raw image, **b** unsharp mask, **c** processed image showing dust shells **d** Upsampled and smoothed final image. 1997 March 1. 30 cm, f/10.5 Newt., Ten stacked 1 second exposures. Nick James.

dynamic range of photographic prints is too small to represent the very wide range of brightness levels found in the average comet. It is much better to scan the original negatives or slides using a film scanner. This maintains the full dynamic range of the film, especially if you can scan with 12-bit depth (4096 grey shades) rather than the normal 8-bits (256 grey shades) (Figure 8.2). Suitable film scanners are available for around £350 and I have used the HP S20 with great success. This scans with a resolution of 2400 dpi which is just right for high resolution films such as Tech Pan and it can produce greyscale depth of 12 bits if you have the right software. If you don't have access to a film scanner you may have some success with the Photo-CD scans which are available at most commercial film processors. The disadvantage of these is that the scan is normally done automatically and comet images tend to

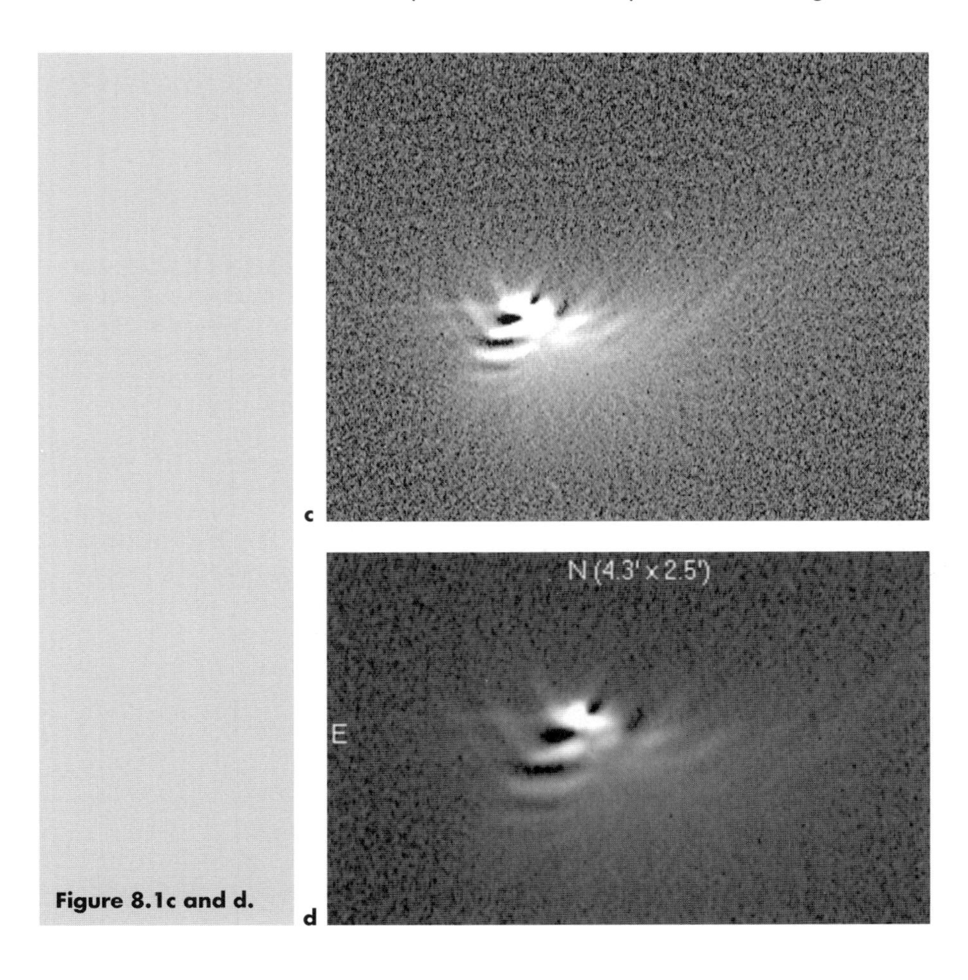

Figure 8.1c and d.

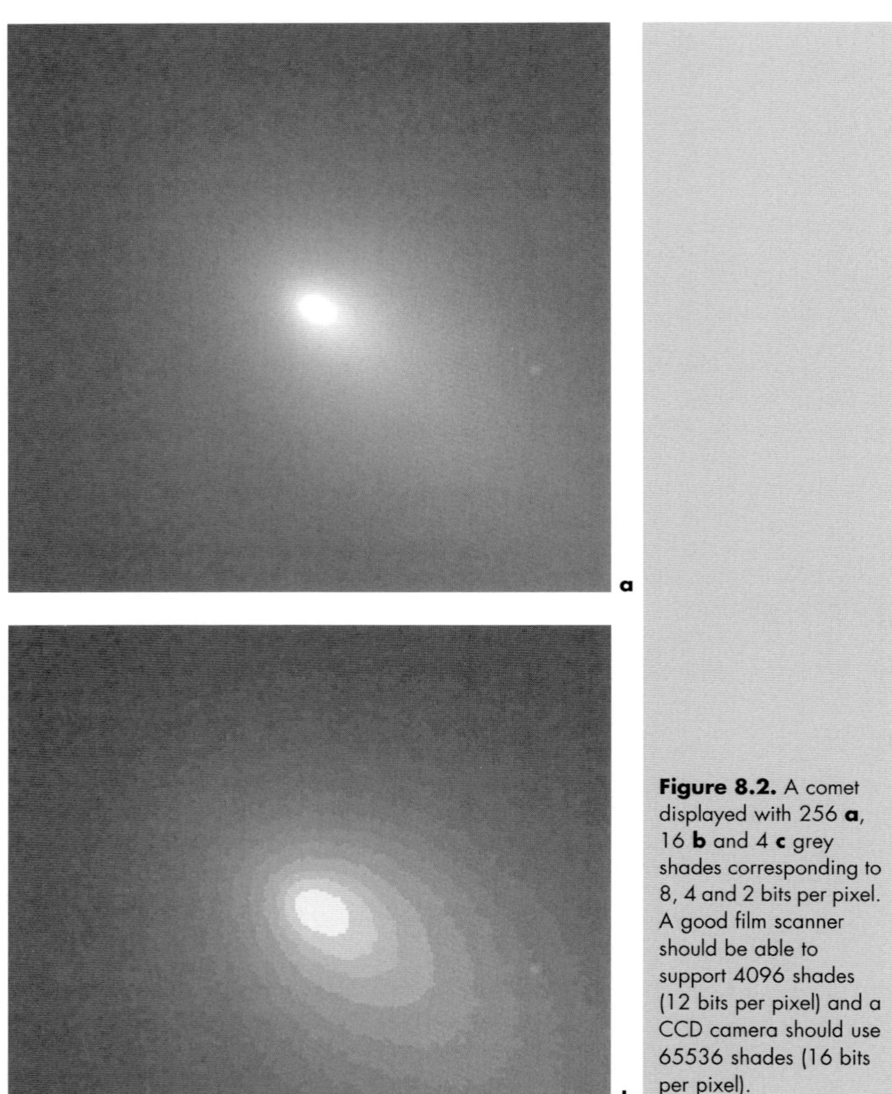

Figure 8.2. A comet displayed with 256 **a**, 16 **b** and 4 **c** grey shades corresponding to 8, 4 and 2 bits per pixel. A good film scanner should be able to support 4096 shades (12 bits per pixel) and a CCD camera should use 65536 shades (16 bits per pixel).

be burnt out on the coma. You will also find that photo-CD scans are only 8-bits deep. Be especially careful that you use true Kodak photo-CD processors. Some cheaper versions only provide compressed JPEG images on the CD rather than full quality TIFFs or PCD files.

Figure 8.2c.

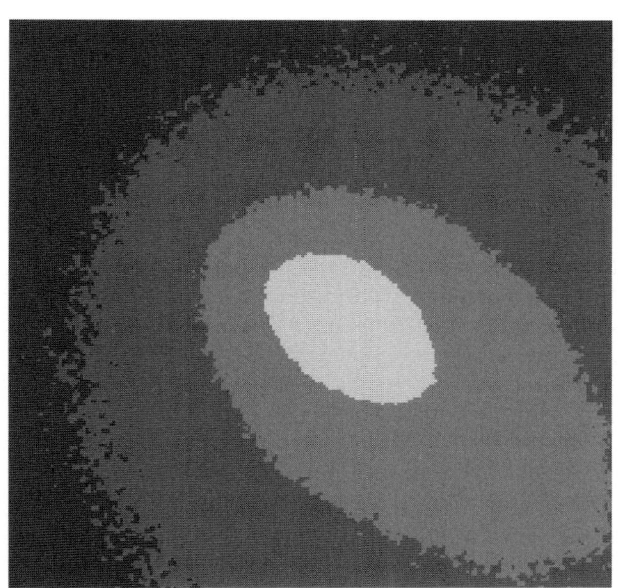

Image File Formats

The standard file format used in astronomical image processing is called FITS (short for Flexible Image Transport System). The large dynamic range of CCDs is lost if you store the images in a file format which does not support 16-bit integer representation. The FITS format has been designed to store astronomical images and so it supports a very wide range of pixel data ranging from 16-bit integers to 32-bit floating point numbers. This latter format supports literally millions of grey shades but in most cases the 65536 shades provided by a 16-bit integer representation will be adequate. This format also stores important information relating to the image such as the time of observation and the length of exposure. FITS is a standardised format which means that, in theory, if your camera software produces a FITS file then it should be possible to read it into any image processing program which supports FITS. Sadly, while some programs and camera manufacturers claim to support FITS they often prefer their own "native" format and, more often than not, they implement the FITS format in a non-standard way. Before you buy a camera or obtain any image processing software you should check carefully with other users to find out how good the FITS support really is. You should use FITS wherever possible, especially if you are collaborating with professional observers.

Commercial graphical file formats such as the Tagged Image File Format (TIFF) are also very flexible but most PC image manipulation programs only understand simple implementations of the format. For example, the TIFF standard allows you to store monochrome images with arbitrary depth but very few programs support anything more than 8-bit greyscale (256 shades) or 24-bit colour (8-bits each for red, green and blue). For comet image processing an 8-bit greyscale is not really enough and so these commercial formats should be avoided if possible.

All is not lost if you have to work with 8-bit input files from, say, a Photo-CD, but you should try to do all of your image processing with a greater depth. The best approach would be to convert the 8-bit files to 16-bit FITS files and then use an image processor intended for scientific or astronomical use. Avoid compressed file formats such as JPEG unless you are preparing an image for submission via e-mail or publication on the web and then only convert to JPEG once you have the final image. The JPEG format reduces the size of images dramatically but there is an irretrievable loss of quality involved in the compression. Many programs have a JPEG "quality" setting so you can trade off image file size against quality but you will notice that pin-point star images don't fare well when subjected to JPEG image compression. At all costs avoid JPEG images as the input for your image processing since information is lost before you even start. Unfortunately, some digital cameras will only provide JPEG compressed images although you can at least select "high quality" compression.

Image Processing Software

Even the most basic image editing software will allow you to improve the dynamic range and detail visible in your final image. If you are simply interested in tidying up images obtained by scanning photographic prints then commercial tools such as JASC's Paintshop Pro, Corel's Photo-Paint, Adobe's Photoshop or any one of a number of different applications will probably do. These applications provide a wide range of tools and they can be used for basic tasks such as changing the contrast and brightness or applying a softening filter or mask to an image. Their support of 16-bit greyscale images is either non-existent (Paintshop) or poor (Photoshop) and so these commercial tools are not particularly suited to more complex image processing tasks. I use them mainly to add captions and other nice features to images that have been processed using more specific tools and to correct for any cosmetic defects in the final image.

For anything other than occasional use you should use a specialised program designed for processing astronomical images. Such programs will handle FITS format files and they will provide a far wider range of routines tailored to astronomical image processing tasks. Some will even allow you to write scripts to automate repetitive tasks or batch process multiple images.

There are many such programs available for Windows. Commercial programs such as MaxIm DL and Astroart (see the Appendix) offer a wide range of features but they cost several hundred dollars/pounds. Other semi-commercial examples include AIP4WIN which comes with the book *A Handbook of Astronomical Image Processing* by Richard Berry and James Burnell. It is also possible to download some excellent programs for free. In my view one of the best image processing programs is Christian Buil's IRIS. This can be downloaded from his website and it can be used for tasks ranging from simple image processing to complex image analysis. I would certainly recommend trying one of the free programs before you consider spending money on any of the commercial packages.

Remember that image files can be quite large and so your PC should have plenty of memory. This is particularly true if you plan to process film scans since a full colour scan of a 35 mm frame at 2400 dpi and 16-bits per colour requires something like 45MB. These large files are not normally a problem for modern PCs and the cost of additional memory or hard disk storage is trivial compared to the cost of a CCD camera or a good film scanner.

Finally, many of the extremely powerful image processing and analysis programs used by professionals are now available to amateurs. Professional workstations of a few years ago have now been eclipsed by the humble PC with its 1GHz processor, 512MB of RAM and 40GB hard disk. Now that the professional's favourite operating system, Unix, runs on PCs we have access to many of their best programs. The ultimate professional package is called IRAF and this runs under Unix operating systems such as Linux on home PCs. IRAF is incredibly powerful but it is also incredibly complicated and not for the faint hearted. Other professional tools, such as the Starlink processing suite have also been ported to Linux PCs (Figure 8.3). The time spent learning these tools is well spent if you intend to do professional level cometary astronomy but they will seem strange to you if your computer experience is limited to the Windows world. The best place to start is to have a look at the various Internet sites suggested in the Appendix.

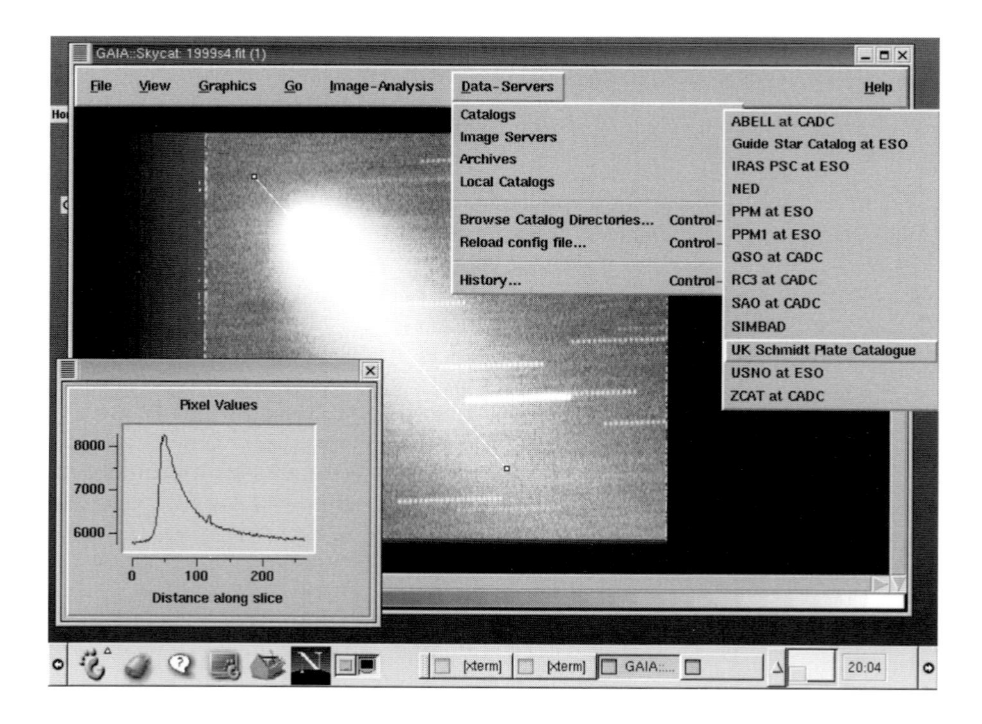

Figure 8.3. A screenshot of Gaia/Skycat, a program distributed as part of the UK Starlink CD-ROM package. This is an example of the software available for Unix-based systems which will run on a home PC. Starlink CD-ROMs are available free for non-commercial use. Gaia was written by Peter Draper and Norman Gray and is based on the ESO Skycat system. Courtesy Starlink.

Basic Image Processing

The first step in the image processing sequence is to apply the calibration frames to your stored comet images. The procedure given here is, as far as possible, independent of any particular image processing program but I suggest that you work through it using a package that you are comfortable with. Some examples of raw image and calibration files in FITS format are included on the CD-ROM. Once you are proficient with the techniques you can process your own comet images and start to extract scientific information from them.

Ideally we would like each pixel in our image to represent light from the sky, comet and background stars with no instrumental effects. In practice the raw image frames are distorted by a number of effects. Each pixel sample normally will have a fixed offset, O, which is due to the design of the camera read-out electronics. The pixel then has a certain amount of thermal noise which is dependent on the exposure time and the temperature of the chip during the exposure. I will denote this part of the signal by $M_r \times D$ where D is a constant for each pixel and M_r varies depending on exposure and temperature. Finally each pixel contains the wanted signal from the comet, P, but this is modified by a pixel dependent sensitivity, S, which varies across the image due to chip effects and variations in transmission through the optical system (dust, vignetting etc.). Mathematically each pixel in the raw image frame, RF, is made up as follows:

$$RF = O + M_r \times D + S \times P$$

To calibrate the raw frame we need a dark frame, DF, a flat field, FF, and a bias frame, BF. The dark frame is taken with no incident light so $P = 0$. The flat field has a uniform illumination so $P = 1$ and such a short exposure that $D = 0$. Finally the bias frame has a short exposure and no incident light so that both $D = 0$ and $P = 0$. The pixels in each of these frames are given by:

$$DF = O + M_d \times D \qquad FF = O + S \qquad BF = O$$

A raw frame of comet C/2001 WM1 (LINEAR) and the corresponding dark frame and flat field are shown in

a

b

Figure 8.4. The raw image file **a**, the dark **b** and the flat **c** associated with an imaging session of Comet C/2001 WM1 LINEAR on 2001 December 1.78. The flat clearly shows shadows caused by dust in the optical path and non-uniform field illumination due to the optics. The dust shadows can also be seen on the raw image along with spurious bright pixels due to thermal noise. These raw images are available on the CD-ROM.

Figure 8.4. Calibration of the raw frame consists of subtracting the bias frame and the corrected dark frame and then dividing by the corrected flat field. Mathematically the calibrated frame, *CF*, is:

$$CF = \frac{(RF - BF) - x(DF - BF)}{(FF - BF)}$$
$$= \frac{S \times P + M_r \times D - x \times M_d \times D}{S}$$
$$= P + \frac{D(M_r - x \times M_d)}{S}$$

The objective is to select the value of x so that $M_r - x \times M_d = 0$ and the last term disappears leaving us with our calibrated pixel value P. Ideally the dark frame (DF) would be taken with an exposure which is the same as the image frame and with the CCD at the same temperature. If this were so then $x = 1$ since $M_r = M_d$ and it is possible to subtract the dark directly from the

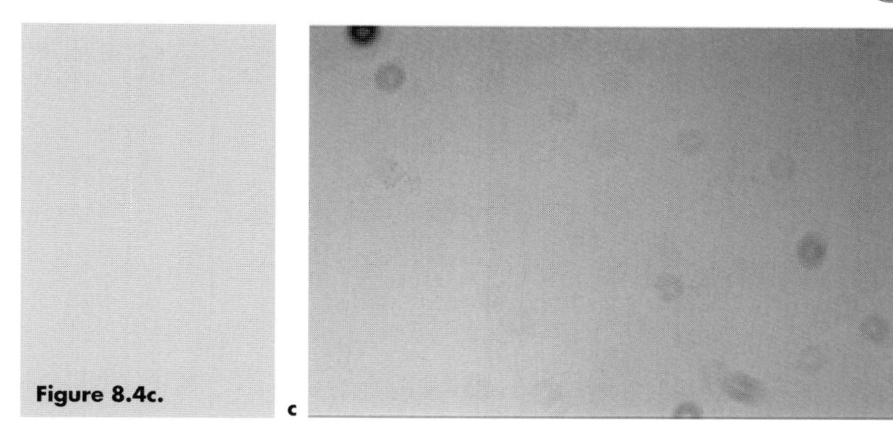
Figure 8.4c.

raw image without worrying about the bias. This can be quite difficult to achieve in practice if the temperature of the chip in your camera is not accurately controlled. An alternative approach is to take a set of dark frames and subtract the bias frame from each. Take the average of the resulting frames to obtain a master dark frame for your camera. This "dark" will be of a very high quality since the averaging process will have reduced the camera read-out noise. You then need to find the value of x which scales this master dark frame to the particular raw image that you are calibrating. Once you have done this, the master dark frame and the bias frame can be subtracted from the raw image to complete the calibration. A number of image processing programs will compute the optimum value of x for you given the master dark frame and your raw image. See the command OPT in IRIS for instance.

If you only have a few image frames to process this calibration can be done manually. Once you have tens or hundreds of frames you will want to do it automatically by applying the same operations to each frame. Many of the better image processing programs allow you to batch process images in this way but if yours doesn't there is a utility on the CD-ROM which you can use.

It can be quite difficult to obtain a good flat field during twilight and the alternative techniques mentioned in Chapter 7 may not be practical either. If this is so you will need to synthesise a flat from existing images. If you image a particularly dull area of the night sky you will find that you have pictures of a few isolated stars separated by large areas of blank sky. If you take many images of slightly different parts of the sky the stars will change position but the flat field variations due to your instrument will stay in the same place. Your computer can then compare each frame pixel by pixel looking for common features (the flat field) and rejecting differences (the few isolated stars). In practice we do this by getting the computer to generate an image where each pixel in the output is the median of all of the corresponding pixels in the input images. Any remaining stars can be removed manually. This is usually achieved by drawing a box around the star and then telling the computer to replace the brightest pixels in the box with the local background (the MAX command does this in IRIS). The resulting image is a good flat field but it does have the disadvantage that there is not much light from the sky on a dark night and so many images are required to get a good signal-to-noise ratio. This is one of those rare occasions where moderate light pollution is an advantage since it increases the signal level from the sky background and it improves the flat field.

It is instructive to compare the quality of the final calibrated image with the original raw image (Figure 8.5). You will probably be amazed at the difference. Accurate calibration using good, high signal-to-noise flat fields and dark frames should allow you to detect features in your images which are as little as 1% as bright as the night sky background. Dust shadows and vignetting can easily cause variations in sensitivity of 20% or more so you can see that a good flat is essential when you are imaging very faint cometary detail. In most types of cometary imaging time spent making the flat is just as well spent as time spent taking the actual exposure.

Figure 8.5. This is the result of calibrating the raw frame (Figure 8.4(a)) using the dark, flat and bias frames. The dust shadows, non-uniform illumination and thermal noise pixels are removed.

More Advanced Processing

Calibration of the raw frames is all that is required if you are intending to use the images for photometry or astrometry. In other cases further processing will be beneficial.

The details in comet images are normally very subtle and so a number of image processing algorithms have been developed which can be used to enhance faint features. It is important to remember that image processing does not generate information, it simply allows you to display images in a way which makes the existing information easier to see. You should bear this in mind when you process your images. There are a great many over-processed comet images submitted to magazines and societies which appear to show impressive features in the coma or tail. Comparison with other images taken at the same time often shows that these features are purely artefacts of the processing applied to the original image. Some of the techniques that I will describe in this section are very powerful but they can be misused and I would recommend caution until you have experience with each method.

The first step is normally to stack the multiple calibrated images that you obtained at the telescope. Before you do this you should reject any frames where the stars are trailed due to drive errors. The stacking should take into account the comet's motion against the background stars and this is quite a time-consuming job if you have to do it manually. Many image processing packages will take a sequence of frames, align them on a reference point and then stack them automatically into a single image. If the comet is well condensed the alignment can be done using false nucleus. If not you can stack using a reference star in the image with a suitable offset between each image to counteract the comet's motion (Figure 8.6). This relies on an accurate time of exposure being available for each image so that the computer can calculate the offset given a PA and rate from the ephemeris.

Comets can have a very large range of brightness levels which are difficult to represent on an output device be it a computer monitor or a photographic print. One of the great advantages of a CCD is that the camera itself has a very large dynamic range which can

Figure 8.6. A total of 14 frames have been aligned on the comet and stacked in this view. The resulting image was then log-stretched to show details in the tail and coma simultaneously. Notice that the noise level is considerably less than that seen in Figure 8.5.

capture all of these details as long as none of the original images are saturated. The problem is how to bring out the details for display. One simple approach is to apply a logarithmic stretch to the original image. The sky background level should be set to zero before you do the stretch. You can then display bright and faint parts of the comet simultaneously and the comet will look similar to the way it does to the human eye. You can also use false colour or brightness contours (isophotes) to achieve the same effect (Figure 8.7) but the image is no longer as easy to interpret.

If you want to reveal small, low contrast features which overlay the general bright background of the coma you can use an unsharp mask. You should first obtain a blurred version of the original image using the "blur" or "low-pass filter" commands in your image processing software. A particularly good filter type to

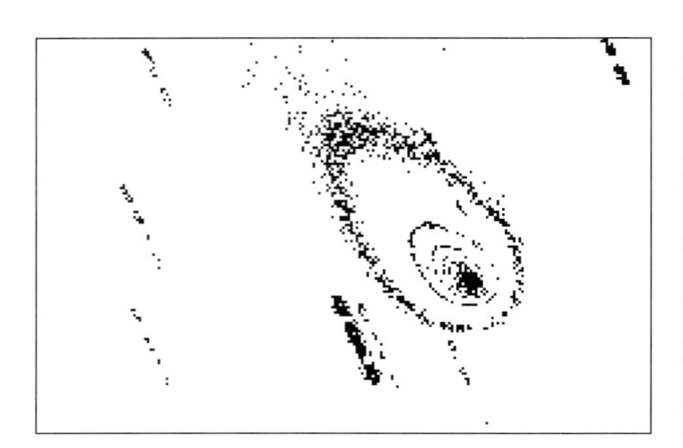

Figure 8.7. An alternative way of displaying the data shown in Figure 8.6. These isophotes represent brightness contours across the comet.

use is a "median filter" since this can completely remove stars from the unsharp image rather than simply blurring them. The original image is then scaled by the unsharp mask. This has the effect of compressing the dynamic range of low frequency detail so that high frequency or small-scale detail shows up much more clearly. This technique was particularly useful with images of Hale-Bopp's dust shells. These shells were very prominent visually in large telescopes but they were superimposed on the bright background of the coma and so they were quite difficult to represent in a computer image. You can see in Figure 8.1(a) that the coma is saturated if the image is printed to show fainter details. The unsharp mask in Figure 8.1(b) was then used to divide the original image and this brings out the shell structure very well (Figure 8.1(c).

Another technique which can be used is called a rotational gradient filter (RGRADIENT in IRIS). You may also see it described as the Larson-Sekanina algorithm. It can be used to isolate low contrast details which are superimposed over a bright object which has symmetry of revolution about a specified centre point. You select the centre point (x,y) (normally the false nucleus) and a rotation angle (θ). It then takes the original image and creates two new images, one rotated by $+\theta$ and the other by $-\theta$ relative to the centre (x,y). These two images are then subtracted from twice the original image. Mathematically, in polar coordinates relative to (x,y), a pixel in the output image R is given by:

$$R(r, \phi) = 2 \cdot I(r, \phi) - I(r, \phi - \theta) - I(r, \phi + \theta)$$

where I is the input image. Any structure which is symmetrical around (x,y) will be removed in the output image and all that is left will be structures, such as jets, which are not symmetrical.

This is a very powerful algorithm but it is also very dangerous in that it can generate apparent radial detail even when none is present. For this reason it should be used with great care and any images submitted which have been processed in this way should clearly state the fact. An example of the technique is shown in Figure 8.8 where a jet near the nucleus of C/2001 WM1 (LINEAR) has been enhanced in this way. In this case the feature was real and the rotational gradient filter brought it out well.

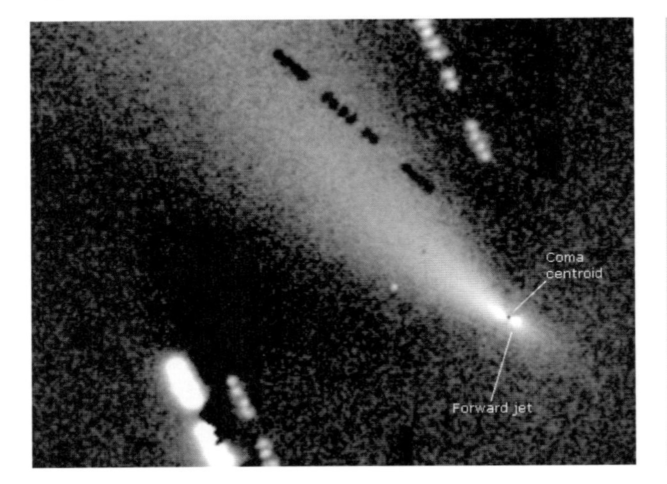

Figure 8.8. This is a magnified version of Figure 8.6. processed using a rotational gradient filter with $\phi = 25°$. It clearly shows a small forward-pointing jet deep within the inner coma.

Colour Image Processing

Colour images of comets can be obtained either from separate monochrome CCD exposures through red, green and blue filters or by scanning colour photographic film. In the case of separate CCD frames the exposure length for the different filters should be adjusted so that the signal-to-noise ratio in each channel is about the same. You can use the catalogued transmission of the individual filters and the response curve of your CCD to calculate the ratio of exposures required for each channel. The sensitivity of the chip and the transmission of the filters means that the exposure in the blue channel is usually much longer than the exposure in the red and green channels. Any departure from the correct ratio of exposures will shift the colour balance but this can be corrected to a certain extent at the processing stage. In fact, it can be quite difficult to get the colour balance right and this is particularly the case for objects such as comets which have strong emission line spectra.

One problem with taking separate red, green and blue exposures is that the comet will be in a different place relative to the star background on each frame. When the images are registered only the comet will be the correct colour. All of the stars will appear in the output image as rows of red, green and blue dots and,

Figure 8.9. Wide-field mosaic image of C/1996 B2 Hyakutake taken on 1996 April 21 using a 180 mm, f/2.8 lens, CCD and narrow-band H_2O^+ filter. Three consecutive frames were taken between $19^h 16^m$ and $19^h 37^m$ UT and each frame was exposed for 3 minutes. The field of view of each individual frame is $10°.7 \times 1°.9$. Herman Mikuz.

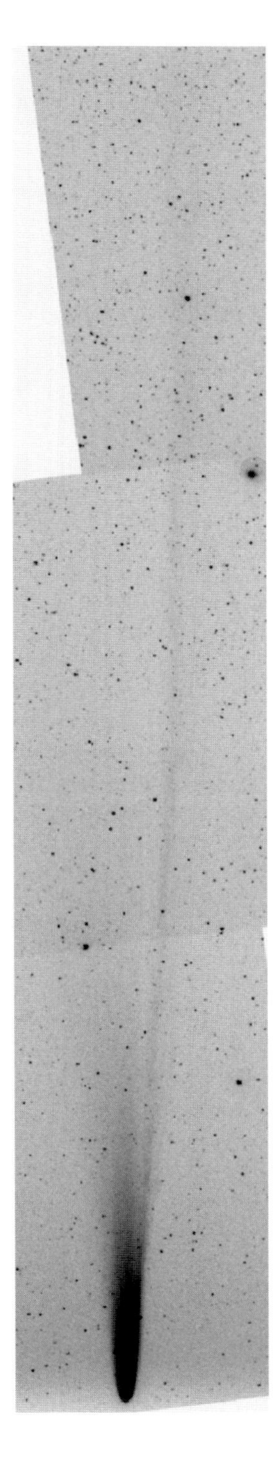

in my view, this rather detracts from the appearance. The only alternative is to take all three colours at once using a one-shot colour camera such as the Starlight Xpress MX7C or go back to using colour film.

The processing algorithms that are used on monochrome images can be applied to colour as well but it is often best to restrict your processing to the luminance, or brightness, component of the image rather than processing the individual colours. Many image editing programs will allow you to transform your RGB image into a format which separates the brightness information from the colour information. One such format is called YUV. The Y, or luminance, channel represents the brightness of the scene and the U and V channels represent what are called colour differences. The Y channel can be processed in the same way that you would process a monochrome image. You can apply unsharp masks, rotational gradients or whatever to bring out the faint cometary detail. This processed luminance image can then be combined with the original colour difference (UV) images to reconstruct a colour RGB image. Processing your images in this way will ensure that you maintain the colour balance and it simplifies the whole procedure since you are only dealing with a single luminance channel.

Specialised Processing Techniques

If a really spectacular comet comes along you will probably find that it will not fit in the field of view of your telescope/CCD camera even with fairly low resolution sampling. One option is to take a set of overlapping images, each with the same exposure, and then mosaic them together. Each image must be calibrated in the normal way and then the sky background should be adjusted so that it is the same in each image. Use your image processing program to register adjacent frames using common stars as reference points. Finally combine the frames into a single image. This must be done carefully or the joins will be visible. Many image processing programs have special commands to do this and there is usually a choice about what to do with the image in the overlap region. A simple average of all the frames in the overlap area usually works quite well. An example mosaic image is shown in Figure 8.9.

If you have multiple frames of a comet you can make an interesting movie rather than just stacking them into a single picture (see the examples on the CD-ROM). Each frame must be individually calibrated and adjusted to a constant sky background level. Unless your tracking is perfect the frames must then be processed so that they are aligned on the comet's nucleus and cropped to the required size. You can then apply the normal algorithms to bring out interesting details and finally contrast stretch as required. After processing each frame should be written out in a standard graphics format such as TIFF which can be handled by your movie encoding software.

To get smooth motion in your movie it must run at somewhere between 15 and 30 frames per second so you will need to obtain and process a large number of frames. It is not normally practical to do this manually and I have written some software which does all of the processing automatically. A copy is on the CD-ROM along with instructions for its use. This produces a sequence of still frames which can be animated using a program such as Animation Shop from JASC. This produces animated GIFs but these are rather limiting and it is much better to encode the frames in a true movie format such as MPEG. A number of freeware and shareware encoders are available and details are given in the Appendix.

Chapter 9

Advanced Research

Every now and again a comet comes along which is bright enough to get public attention. Such comets are the exception rather than the rule and you could get the impression that cometary astronomy consists of short periods of excitement separated by long periods of inactivity. Nothing could be further from the truth. At most times there are several comets available for observation. These comets are usually faint and most of them don't have tails but there is plenty of work to be done by amateurs equipped with moderate-sized telescopes and CCD cameras. Accurate positions can be obtained using astrometry and the comet's brightness can be recorded using photometric techniques. Routine monitoring can sometimes show sudden, unexpected changes in brightness which occur when the comet becomes more or less active. These faint comets are the preserve of a few dedicated comet observers but modern equipment means that they are much easier to observe than they used to be.

If you follow the suggestions in this book there is no reason why you cannot produce scientifically useful results when you observe comets. Some people say that "doing science" is boring and detracts from the enjoyment of observing. I say exactly the opposite! I can still gaze in awe at a comet but my observing enjoyment is enhanced dramatically by the knowledge that I am making a contribution, no matter how small, to the study of these fascinating objects.

Astrometry

It is very important to define the orbit of a comet as soon as possible after its discovery. Thereafter it is necessary to continue to refine the orbit so that we can improve our knowledge of the comet's history and make accurate predictions about where it will be seen in the future. With the increasing incidence of spacecraft being sent to study comets the ability to predict accurately where a comet will be at some future date becomes even more critical. Finally, in very rare instances, it may be that a comet makes a close fly-by of the Earth so there might be some risk of a collision in the future. It is only by accurately characterising the orbit that the level of risk can be determined.

To determine the orbit of a comet is all we need is its position and velocity at a given time. In practice it is very difficult to measure these quantities directly and almost all orbits are computed from what are called *astrometric* positions. These are very accurate measurements made of the comet's position against the star background. Three well-spaced astrometric positions are usually enough to give an initial orbit. Further observation will improve the orbit over time.

The basic technique is to obtain an image with an exposure which is just long enough to obtain a good signal-to-noise ratio at the brightest part of the coma. The UTC times of the start and end of the exposure should be recorded to an accuracy of at least one second. The position of the comet is then obtained by measuring its X–Y coordinates on the image along with the coordinates of a number of *reference stars*. These stars are selected since they have accurately known positions taken from an astrometric catalogue and they can be used to determine what are known as the *plate constants* of the system. These constants define how the X–Y coordinates of objects in the image relate to real RA and Dec positions on the sky and so they can be used, along with the measured coordinates, to determine the comet's position.

Up to about twenty years ago astrometric catalogues were very sparse and so very wide fields were required to ensure that the comet and a few catalogue stars could be recorded on the same image. The recording medium was usually a large photographic plate and the X–Y coordinates of the comet and stars were measured using a measuring engine. Measuring engines ranged in size from small, home-made, table-top devices to huge beasts the size of a filing cabinet. In all cases the objective was to measure the linear position of an object on the film or plate to an accuracy of a few microns. The observer would carefully measure the positions of the reference stars and the comet using a microscope eyepiece and would then extract the star positions from a printed catalogue. He would then reduce the results to get the actual RA and Dec position of the comet. This process was painfully slow and complicated and only a very few observers had the necessary skills to do it accurately.

In the late eighties two things happened that dramatically changed the way astrometry was performed. At around this time CCD cameras had started to appear in the amateur community. CCD chips are ideal for astrometry since they consist of a regular array of pixels which provide a very accurate reference grid. Furthermore computer programs can automatically extract the X–Y coordinates of objects to an accuracy of a small fraction of a pixel. This immediately eliminated the arduous job of measuring each star using a mechanical measuring engine. On their own the appearance of CCD cameras would not have been sufficient. The CCDs available to amateurs had small fields of view such that very few (or no) stars from the conventional astrometric catalogues would be available for measurement. The second important factor was the availability of star catalogues containing a far denser grid of stars. In 1989 the first edition of the Hubble Guide Star Catalogue (GSC) was published. This contained reasonably accurate positions for stars down to 16th magnitude and so plenty of reference stars were available even in narrow fields. Initially this data was quite difficult to use but in 1992 a revised version of the catalogue came out on CD-ROM. From that time onwards astrometry became a straightforward process and these days computer tools exist which automate the entire reduction.

Practically anyone with a CCD camera and a PC can now produce very high quality astrometry with positional errors of well under an arcsecond. For the best results your image scale should not exceed 2 arcseconds per pixel and the start and end time of the exposure needs to be recorded accurately (to better than 0.00001 day, or about 1 second). Accurate timing is important if you want to measure a comet's position with an accuracy of tenths of a arcsecond. It only needs to be moving at a few hundred arcseconds per hour for timing inaccuracies of a second or two to become significant.

PC clocks are notoriously unreliable and drifts of 30 seconds a day are common so you should make sure that you set your PC's clock accurately before starting an observing session and check it again afterwards. An even better solution is to arrange for your PC to be automatically synchronised to UTC. This can be achieved with add-in time signal cards but these can be expensive. A much cheaper alternative if you have an Internet connection is to use a network timeserver. The Network Time Protocol (NTP) can keep your computer synchronised to within a few milliseconds of true UTC under Unix systems. A simpler version of the protocol (Simple NTP) can be used with Windows and there are a number of free utilities available for download (see the Appendix).

Once you have the images any modern astronomical image processing program will perform the reduction automatically. An example reduction is shown in Figure 9.1. In the first step the software identifies stars and objects in the image. A centroiding algorithm is used to obtain the X–Y centres of each object to a fraction of a pixel and the identified stars are compared against a catalogue to look for common objects. In order for this comparison to work you normally need to tell the software the image scale and an approximate position for the centre of the frame. Once stars on your image have been matched with stars in the catalogue the software computes the plate constants and then the astrometric position of the comet from its X–Y coordinates in the image. The time reported should always be the mid-time of the exposure since, even if the comet is trailed during the exposure time, the centroid algorithm will select the middle of the trail. A computer is also very good at determining the centroid of objects which are trailed due to guiding errors. The computer, unlike a human observer, will consistently choose the centroid at the same point so that the errors are minimised.

Most astrometric software will directly produce data in a format which is suitable for e-mailing to the Minor Planet Center (MPC). An example of this format is shown in Chapter 6. When you report positions you should include your observatory's latitude, longitude and height above sea level to an accuracy of better than 100 m. If your astrometry is of good quality the MPC will allocate you a three-figure observatory code which can be used in future reports.

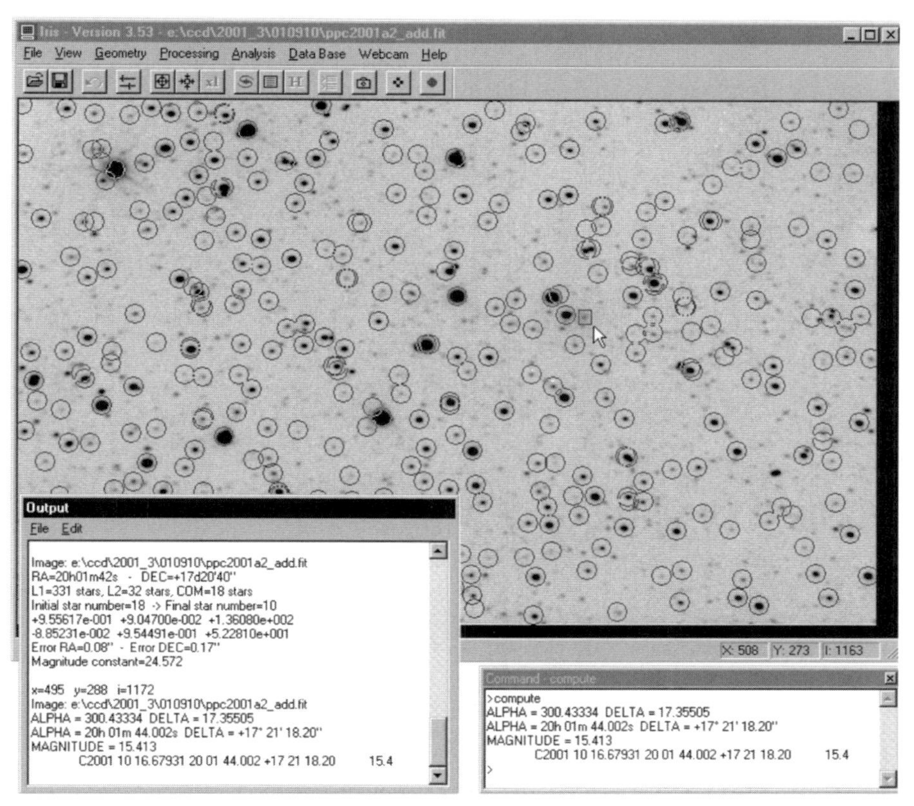

Figure 9.1. Automatic astrometry of C/2001 A2 LINEAR (compare Figure 7.11) using IRIS. The program has detected 331 stars in the image and 32 stars in the source catalogue (GSC-ACT). It then determines that there are 18 common stars which it proceeds to use in the reduction. When the reduction is complete 8 of the stars are rejected and the final plate constants are determined by 10 stars. The position of the comet (in the square box) is then computed. The magnitude given is based on GSC magnitudes which are not accurate. IRIS software courtesy Christian Buil.

Despite the fact that astrometry is automated there are a few things to watch out for. The centroiding algorithm in your software should be able to determine the X–Y position of stars to around 10% of the pixel size and so the astrometry should be accurate to a few tenths of a second of arc at image scales of 1–2 arcseconds per pixel. In most cases the accuracy is limited by the catalogue positions of the reference stars. The original GSC was not particularly good but it has now been re-reduced using more modern astrometric data to produce a catalogue called the GSC-ACT. Programs for converting GSC V1.1 CD-ROMs to GSC-ACT are available online at Project Pluto's website and they can also supply the data in CD-ROM form. A denser catalogue, the USNO-SA2.0, is also available on CD-ROM. It contains over 54 million objects with an accuracy similar to the GSC-ACT. Recently an even better catalogue has been announced. The US Naval Observatory CCD Astrograph Catalog (UCAC) will contain 2000 stars per square degree with a much better accuracy than the GSC-ACT. It should have complete sky coverage by 2003.

Another pitfall is that the computer will make its own estimate of the centroid of each object on the image. This works well for stars which should have a regular shape but the optical centroid of the comet's coma may not be the actual location of the nucleus. Jets and other activity will tend to skew the central condensation and this can make it difficult to determine the actual nucleus position. It is best to use as small an aperture as possible and in all cases the computer's determination of the comet's centroid should be checked to ensure that it looks reasonable.

When performing astrometric reductions you should be careful that you don't rely entirely on the computer to get things right. Don't forget that it is simply a tool and you are responsible for ensuring the accuracy of the results you obtain. Before submitting your observations to the MPC it is a good idea to check them against the known orbit or get someone else in a local group or national organisation to do this for you. Since the comet may be some way off its predicted orbit you should look at the consistency of the errors you get for multiple measures made on the same night rather than at their absolute size.

Cometary Photometry

Unlike astrometry, which has been made reasonably straightforward by modern software packages, photometry, or the measurement of a comet's brightness, is still difficult to get right. For a start it is necessary to define what we mean by the brightness of a comet. Comets are extended objects which have a wide range of surface brightness values ranging from the centre of the coma to the extremities of the tail. The *total magnitude* of a comet, denoted m_1, is supposed to be the integrated light received from the entire coma but excluding the tail. Measuring this is difficult in practice since the coma gradually merges with the sky background and it is often overlaid by brighter parts of the tail.

A second problem involves how different cameras respond to the light from the comet. The main emission from faint comets comes from diatomic molecules of carbon, C_2. This molecule has several emission lines over the visual spectrum in groups called the Swan bands. The main lines are at 515 nm and 560 nm but other lines exist. If you look back at the response curves of common CCD cameras (Chapter 5) you will see that they all have a different relative response to different colours of light. The result of this is that you would get a different magnitude depending on which camera you use. This is obviously unsatisfactory and to avoid it you must ensure that your camera responds to light in a defined way. This is achieved by using coloured filters in front of the CCD so that the combined response of the chip and the filter approximates to a standard *photometric band*. The filters to use for any particular photometric band should be recommended by the manufacturer or designer of your camera.

The standard colour bands are called UBVR and I for ultraviolet, blue, visual, red and infra-red (Figure 9.2). V-band photometry is the most common since this approximates to response of the dark-adapted eye. The

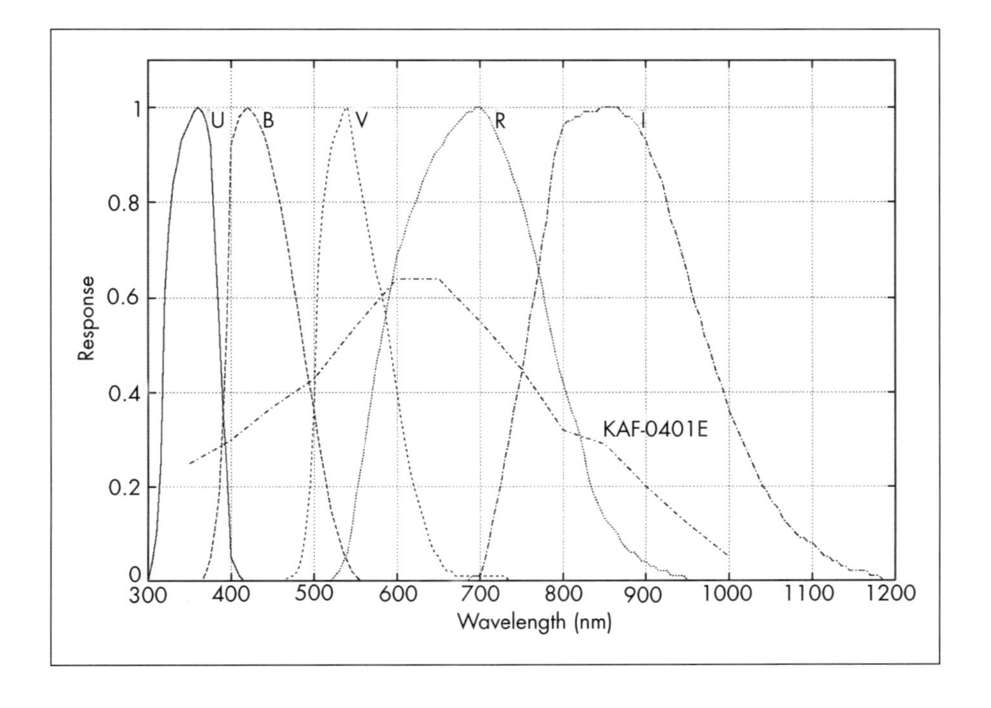

Figure 9.2. The passbands of standard Johnson U, V, B, R and I filters are used to define the response of photometric equipment so that observations between observers can be compared. The response of a KAF-0401E CCD is also shown for comparison.

extended red response of CCD cameras makes them particularly suitable for R-band photometry as well. In normal circumstances these filters are mounted in a filter wheel so that different ones can be placed in the light path during an observing session. Remember to take flat fields for each filter that you use so that the images in each band can be calibrated.

With care accurate photometry can now be obtained with relatively small instruments. In fact, for really large comets you will need to use short-focus lenses to get all of the coma in. Remember that the field of view must be large enough to include the entire coma and sufficient surrounding sky to make an accurate estimate of the background sky level. There is no need for fine image sampling and it is often preferable to bin your CCD to obtain better signal to noise ratio especially if the pixels are small. Scales of 5 to 10 arcseconds per pixel would be suitable in most cases.

The best technique is based on successive imaging of the comet and then a comparison star of known magnitude. It is very rare to be able to find a suitable comparison star in the same field as the comet so what you should do is choose a comparison star at around the same altitude and as close to the comet as possible. This will avoid effects due to differential atmospheric extinction. The comparison star magnitude and colour should be well known. Usually the Tycho catalogue can be used for brighter stars. Take an exposure of the comet which is long enough to record a significant part of the coma, but not so long as to saturate on the false nucleus, then move off to image the comparison star. Since the star is a point source it will require a far shorter exposure than the comet. As always make sure that you do not saturate the star. Repeat this several times slewing back and forth between the comet and the star. For brighter comets you may need to defocus the telescope slightly so that the nucleus doesn't saturate while still bringing out the full coma. If you do this make sure that you image the star at the same focus setting as you use for the comet. Slight defocusing can actually improve accuracy if the image is under-sampled since it will spread the light from the comparison star over several pixels.

Before you analyse the results make sure that the frames are accurately calibrated using the techniques described in Chapters 7 and 8. If you want to get accurate photometry it is particularly important that you use good flat fields since the sky background estimate will have a direct bearing on the results. You can then measure the comparison star on each of the reference frames. Image analysis software will allow you to place an aperture around the star (Figure 9.3). It will then compute the total counts in this aperture and subtract an estimate of the sky background. Measuring the star is relatively straightforward. If you have well exposed images you should find that the counts you get for the star do not vary by more than a few percent.

Measuring the frames containing the comet is much more difficult. The first thing that you must do is remove any stars which are seen within the comet's coma. Most programs allow you to place a small box around each star and then remove it by replacing the pixels in the box with pixels corresponding to the immediate surroundings (see the command MAX in IRIS) (Figure 9.4). Once you have removed the stars you must determine where the coma ends. You can then define an aperture which encloses the complete coma and ask the program to compute the counts in this aperture. To do this it must make a very accurate estimate of the sky background level and this is why you should arrange for the comet to only fill a relatively small part of the field of view. To determine where the coma ends you can experiment with larger and larger apertures until the count does not increase by much.

Figure 9.3. Software can form an aperture around the target object and then count the light intensity from all of the pixels within the circle. This example of C/1999 S4 LINEAR on 2000 July 19 shows the aperture centred on the coma. The median sky background in the frame has been set to zero. Photometry in IRIS screenshot.

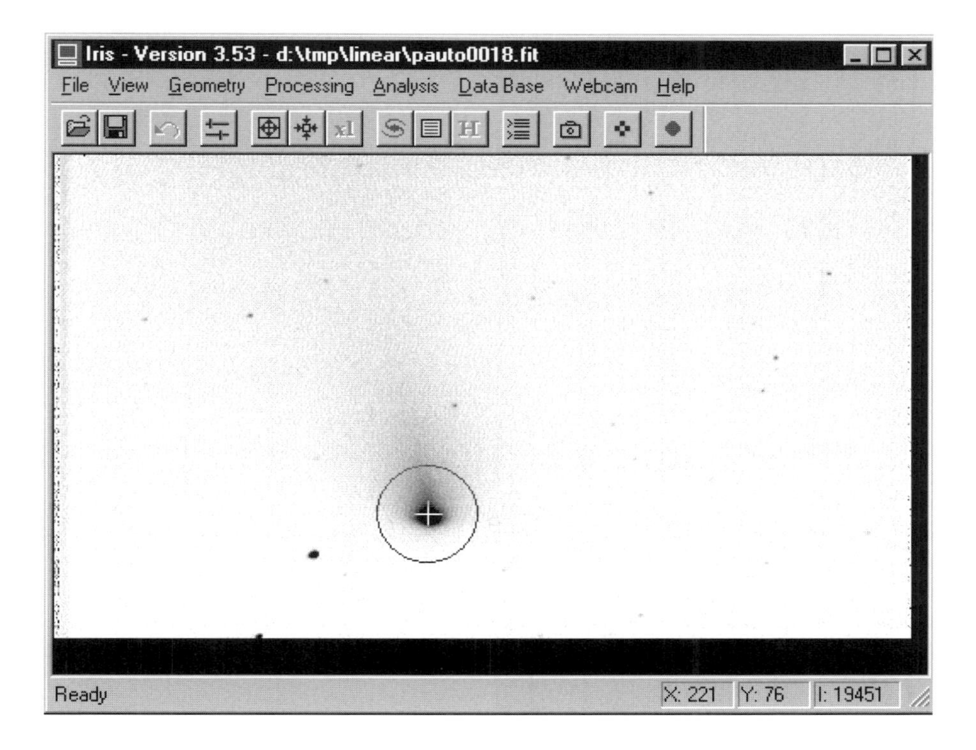

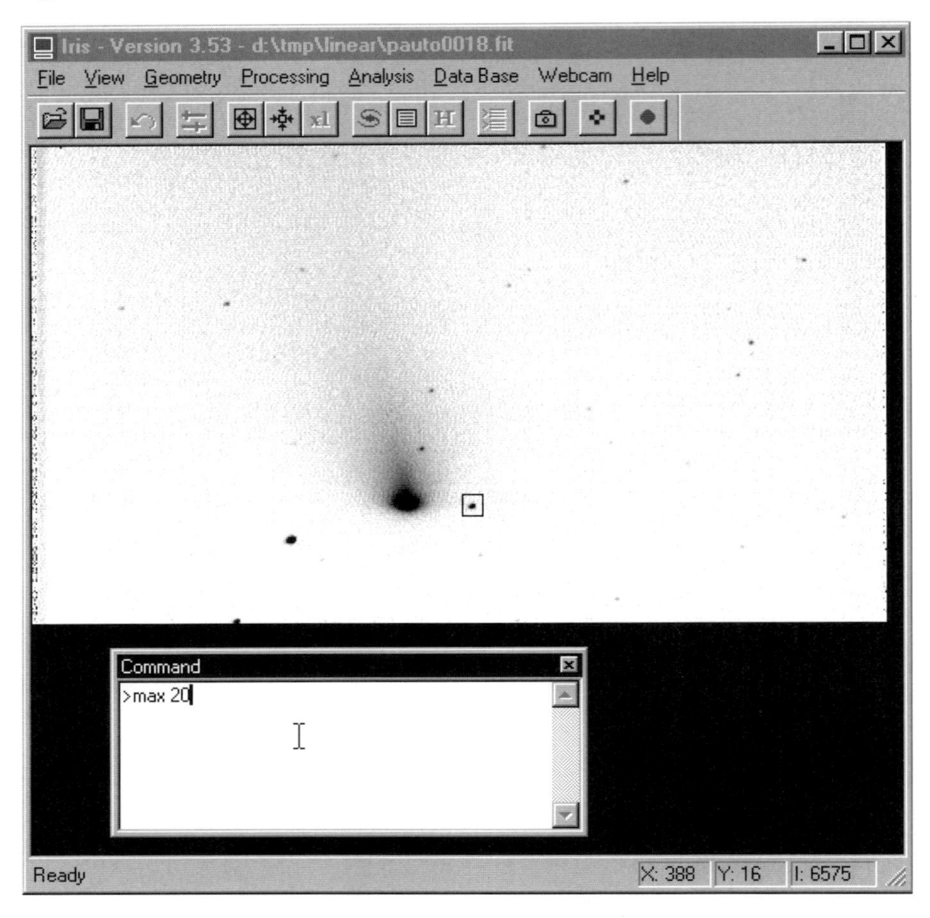

Figure 9.4. Some of the stars on the original image used in Figure 9.3 look as if they could cause trouble during the photometric reduction. In this screenshot one of the offending stars has been identified and it will be removed using the IRIS MAX command. IRIS software courtesy Christian Buil.

Once you have a count for the comet and a count for the star you can determine the comet's total magnitude, m_1, as follows:

$$m_1 = m_s - 2.5 \cdot \log_{10}\left(\frac{C_c}{E_c} \cdot \frac{E_s}{C_s}\right)$$

where C_c, C_s are the counts for the comet and star respectively, E_c and E_s are the exposure durations for the comet and star and m_s is the catalogue magnitude of the comparison star in the photometric band that you are using. This assumes that the comet and star are both near the same altitude. Light coming from an object near the horizon has a longer path through the atmosphere than light from an object near the zenith so two objects of equal brightness but different altitude would appear to have different magnitudes (Figure 9.5).

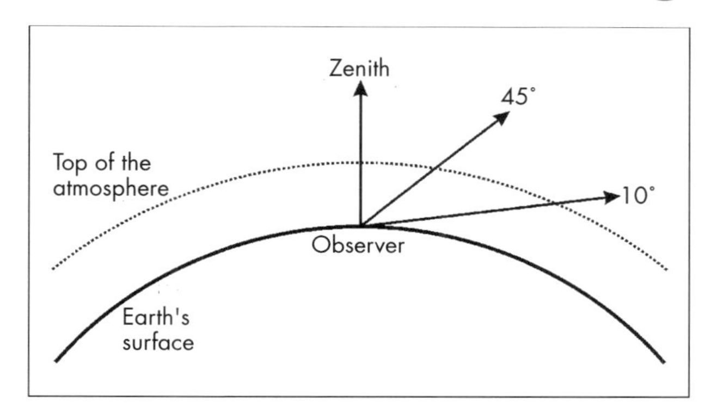

Figure 9.5. Light from stars near to the zenith travels through less atmosphere than light from stars lower down. Cometary photometry should be corrected for this effect if the comet and comparison stars are at significantly different altitudes.

This effect is called *atmospheric extinction* and it can be a serious problem since comets are often observed when they are near the horizon. A handy empirical formula for the extinction, m_e, was derived by Georgii Rozenberg:

$$m_e = \frac{A}{\cos \xi + 0.025 \cdot \exp(-11 \cdot \cos \xi)}$$

where m_e is the extinction in magnitudes and ξ is the zenith distance (90° minus the altitude). A is a constant

Figure 9.6. Plot of atmospheric extinction against altitude calculated using the formula in the text. The upper curve assumes A = 0.3, the lower curve A = 0.2.

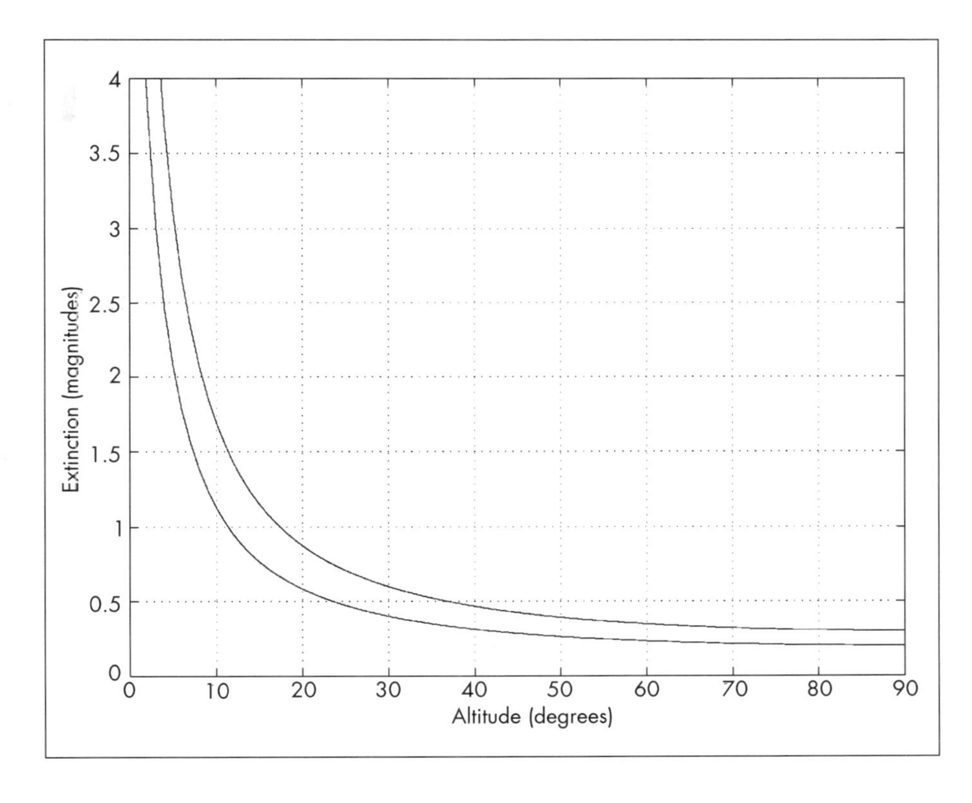

which is dependent on your observing site and the conditions. At a good dry site A is around 0.2, at a damp sea-level site it may be 0.3 or more. A plot of extinction against altitude is shown in Figure 9.6 and you can see that the corrections are significant when you are observing objects below around 30°.

Photometry of comets is an important subject which is neglected by most observers. An accurate light curve tells us a lot about the chemistry and physics of a comet and your photometry could make a real contribution to cometary science. You should try to follow the comet for as long as possible after perihelion since the most interesting things will happen when everyone else has stopped looking.

Cometary Outbursts

It is quite common for a comet to suddenly flare up and become much brighter than predicted. Outbursts in bright comets are often detected rapidly since there are many people observing them but outbursts in fainter comets can be missed. These outbursts, which often occur at large heliocentric distances, are important in understanding the chemistry of comets.

The best example of a faint comet prone to outbursts is comet 29P/Schwassmann–Wachmann 1 which has an almost circular orbit about 6AU from the Sun. Normally this comet varies between magnitude 17 and 19 but every now and again it has an outburst

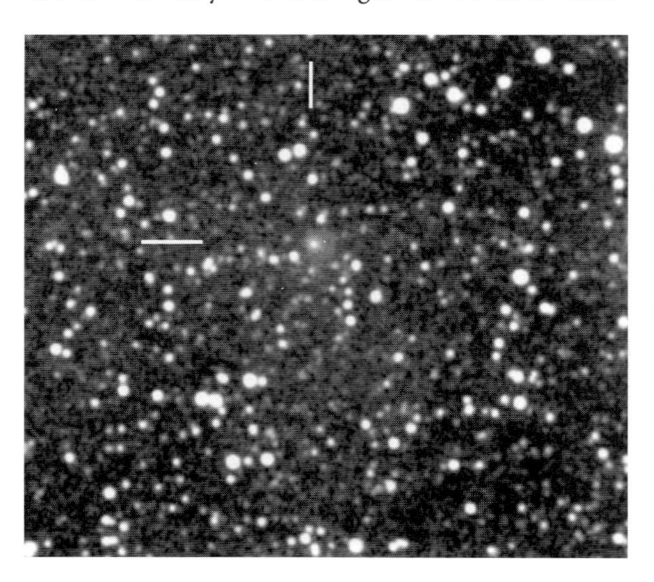

Figure 9.7. Comet 29P/Schwassmann–Wachmann 1 imaged during an outburst. This image consists of 15 co-added 30 sec frames obtained on 2001 June 20 using a Hi-Sys24 CCD attached to a 30 cm, f/2.8 Baker–Schmidt camera. G. Sostero and P. Blasich.

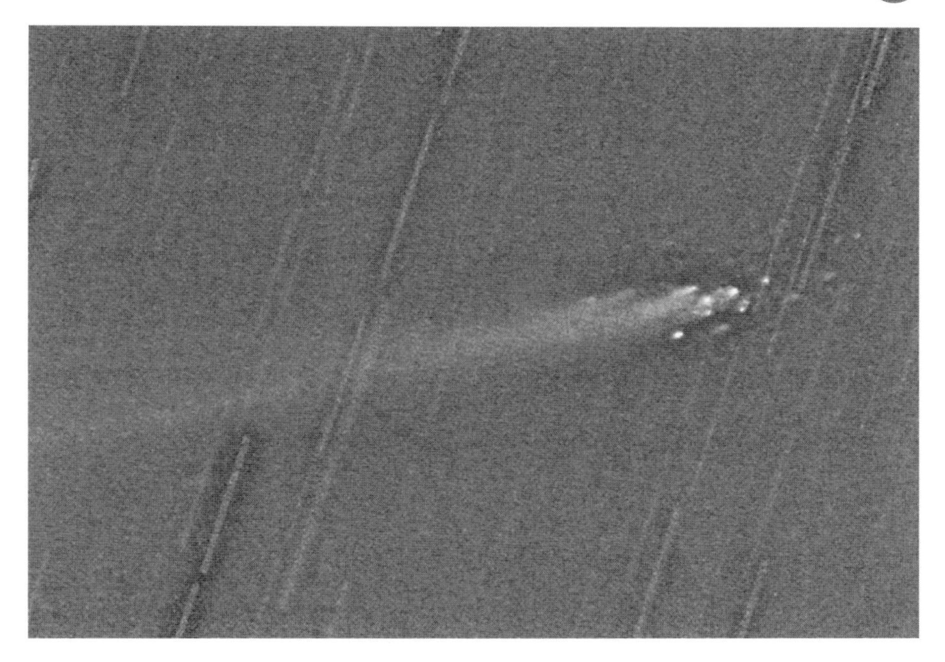

Figure 9.8. Comet C/1999 S4 (LINEAR) had been a normal comet up until 2000 July 25 when it was seen to split into multiple pieces. This image was obtained using the ESO Very Large Telescope on 2000 August 6. More than a dozen "mini-comets" can be seen in this unsharp masked image. Field of view is 3.4 × 2.5 arcminutes. North up. Image processing by Richard Hook (ST/ECF, Garching).
Courtesy European Southern Observatory.

taking it to magnitude 12 or so (Figure 9.7). There are other comets which do similar things but very few people keep an eye on them. A regular monitoring program using CCD photometry and a moderate telescope would be worthwhile.

Faint comets are often under-observed but they have been known to split into pieces. A good example was seen recently when the nucleus of Comet 51P/Harrington split. This comet had come to perihelion in June 2001 but towards the end of 2001 it had faded to around magnitude 20. In December an observer using a 30 cm, f/6.3 SCT detected it at magnitude 17. It had split into two parts and subsequent observations showed the two components drifting apart. Other comets have done the same (Figure 9.8) and this just shows how important it is to keep a comet under observation when everyone else has stopped looking.

Comet Spectroscopy

The only way that we have of directly sensing what a comet is made of is to take a spectrum of it and see how its light is made up. The basic concept of a spectroscope is straightforward (Figure 9.9). Light from the telescope passes through a narrow slit at the focal plane to isolate the particular object of interest. The light from the slit is this then collimated into a parallel beam using a suitable lens. This beam is then split into its component colours by a prism or diffraction grating and an image of the slit for each separate colour is transferred to the detector (usually a CCD) by another lens. The resolution of the spectrum is determined by the pixel size of the detector and certain parameters of the spectrograph (see *Advanced Amateur Astronomy*, Chapter 15, for a full discussion).

In practice spectroscopy of faint comets is difficult since the light from a small part of the comet is spread out over the CCD to form a spectrum. Amateur observers have been successful with brighter comets (Figure 9.10) and it is fascinating to be able to directly sense what a comet is made of using equipment in your own back yard.

There is not enough space here for a detailed discussion of spectroscopy but there are number of resources listed in the Appendix if you want to have a go yourself. A number of commercially built spectroscopes are available on the market or you can build one yourself.

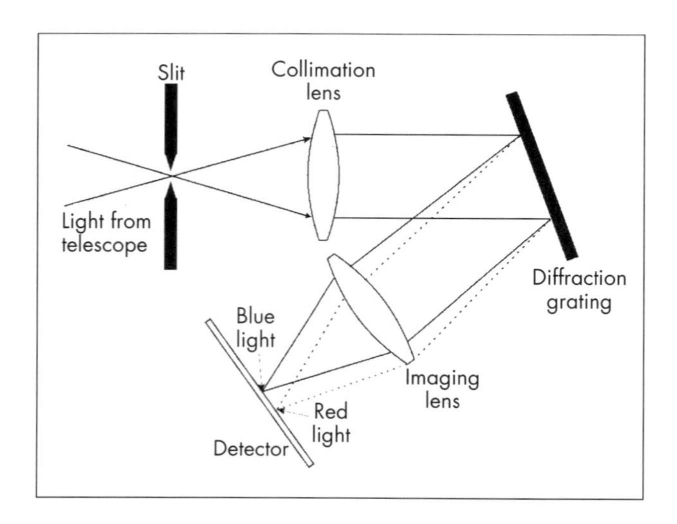

Figure 9.9. A simple spectroscope.

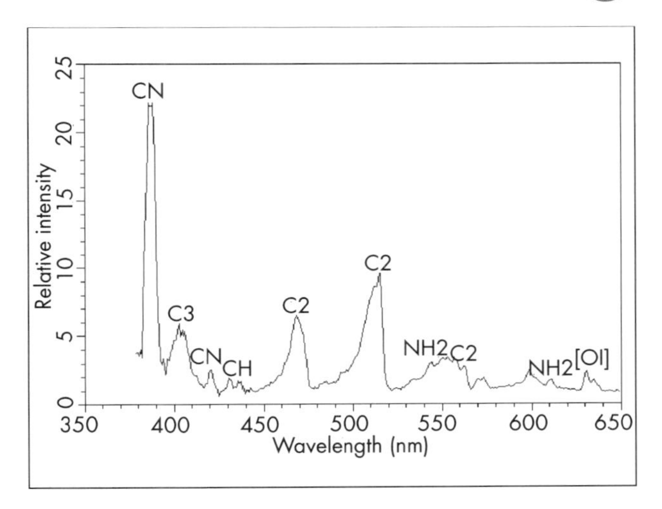

Figure 9.10. Spectrum of the inner coma of C/2001 A2 LINEAR on 2001 July 11 obtained using a Spectr'aude spectrograph, an FSQ-106 Takahashi refractor and an Audine KAF-0401E CCD camera. This spectrum is corrected for the varying CCD response and the sky background has been subtracted. The Swan Bands due to carbon (C_2 and C_3) are clearly visible as is a large component due to cyanogen (CN). Christian Buil.

Comet Discovery

Most astronomers dream of discovering a comet. Comets are the only astronomical objects that carry the discoverer's name and they confer a particular kind of immortality that is not available elsewhere. If that wasn't enough then amateur discoverers have a further incentive in the Edgar Wilson award. This is worth $20K per year and it is awarded exclusively to amateur comet discoverers. Discoveries can be visual, photographic or electronic but they must be made with privately-owned equipment.

The amazing British observer George Alcock discovered a total of five comets using nothing more sophisticated than 25 × 105 binoculars. His last comet was discovered in 1983. At that time the field was wide open to amateur observers but in the mid-nineties professional astronomers started to search for potentially hazardous Near Earth Objects (NEOs) using large telescopes, huge CCDs and prodigious amounts of computer power. As a by-product these automated searches began to pick up practically all of the comets that swept through the inner Solar System.

The latest professional comet-hunting programs are even more voracious and very few comets are left for the amateur to discover. The Lincoln Laboratory Near Earth (LINEAR) search programme, for instance, uses 1m telescopes designed for the US Air Force in conjunction with 2560 × 1960 large format frame-transfer CCDs. With 100 second exposures their system reaches magnitude 22 over a 2° square field of view and the frames are automatically searched for new objects using a computer. How can we possibly compete with that?

A brief glance at the statistics shows that the number of comets discovered by amateurs (discounting the SOHO comets discussed in Chapter 6) has dropped dramatically in the past few years. There are still some amateur discoveries but they are mainly from the southern hemisphere. Syogo Utsunomiya and Albert Jones, observing from Japan and New Zealand respectively, discovered C/2000 W1 using binoculars and a small telescope when it was in the far south. In 2001 there were 37 comet discoveries but only one of these was an amateur discovery and that was a fluke. Vance Petriew discovered C/2001 Q2 using a 20-inch reflector at a Star Party while star-hopping across Taurus to the Crab Nebula! The fact that he could do this from Canada shows that the automatic searches aren't yet infallible and, amazingly enough, 2002 saw a sudden increase in amateur discoveries with three comets (C/2002 C1 Ikeya-Zhang, C/2002 E2 Snyder-Murakami and C/2002 F1 Utsonomiya being discovered in quick succession. All of these comets had been missed by the automatic searches so there is still some hope for potential amateur discoverers.

Even though it is extremely unlikely you may one day come across a comet that has escaped the automated searches it is best to know what to do if this happens. The first thing to do is to confirm that the object is real and that it shows motion. "Ghost images" of nearby bright stars are common in optical systems and can often be mistaken for fuzzy comets. Comets should be seen to move over periods of a few minutes to an hour or so. If it doesn't move it probably isn't a comet! Use a good atlas or catalogue to check that your "comet" isn't a cluster or galaxy and use a computer service such as the one operated by the Central Bureau for Astronomical Telegrams (see Appendix) to check that it isn't a known object.

If you are convinced that you have a new comet measure the position as accurately as you can. Visual observers should be able to get a position accurate to better than a minute of arc. CCD observers should be able to get sub arcsecond accuracy. Make sure too that you report an accurate time (in UTC, not local time) along with a description of what you have found.

You can report your discovery potential directly to the Central Bureau but it is often better to report via a national organisation such as TA in the UK or ALPO in the US. These organisations have groups of observers who will check out your claim before passing it on to the professional body. In all cases you should include your full name, phone number, e-mail address and details about your observing location and the instrument you used. Most importantly you should report which sources (atlases, charts) you used to check out the object.

Discovering comets was always a matter of great dedication or luck but it has now been made much more difficult by the automated professional searches. The two great comets seen at the end of the last century were both discovered by amateurs just before the big automated searches really got going. Both Hyakutake and Hale–Bopp put on a magnificent show and they taught us a lot about comets. They are the subject of the final chapter.

Chapter 10 Recent Great Comets

Great comets are rather like buses. You can wait for years for one to arrive and then two come along at once. This is what happened in 1996 and 1997 when we experienced two comets worthy of the "Great" title. Comet Hyakutake was discovered only two months before it reached its best. It was a small comet which came very close to the Earth and gave us a spectacular view for a few short weeks. Comet Hale–Bopp on the other hand was a leviathon. It was discovered a full 18 months before it was at its best and, during the winter and early spring of 1997, the comet dominated the northern sky. Since it was so bright for so long practically everyone, from dedicated comet observers to even the most disinterested member of the public, saw it. Hale–Bopp was probably seen by more people than any other comet in history.

No one knows when the next Great Comet will come along. The lessons that we learnt with Hale–Bopp and Hyakutake will be useful whenever the next spectacular comet is discovered be it next week or in twenty years' time.

The Discovery of Hyakutake's Second Comet

On 30 January 1996 the Japanese amateur astronomer Yuji Hyakutake left his house in the village of Hayato, about 600 miles south-west of Tokyo, and set off for his normal observing site. After a journey of around 10 miles he stopped and set up his equipment ready for the night's work. Since the previous July he had been avidly searching for comets using 25 × 150 binoculars. These had already proven themselves a month earlier when he had discovered the first Comet Hyakutake. That comet wasn't particularly bright but he was planning to take some photos to record its progress. Frustratingly, the patch of sky where his comet would be was obscured behind some cloud so he started to scan the clear areas using his binoculars. Just before 5am, as he was sweeping through Libra, he noticed the 11th magnitude fuzzy object which was to make him famous throughout the world. The comet was officially named C/1996 B2 (Hyakutake) but at that time there was no indication that this was anything other than a run-of-the-mill, faint comet.

CCD astrometry of the comet quickly followed Hyakutake's discovery and by 3 February the first parabolic orbital elements were reported. These showed that the comet would reach perihelion on 2 May at a distance of only 0.22 AU from the Sun. A close perihelion distance was remarkable enough but the elements also indicated that the comet would approach to within 16 million km of the Earth in late March. Hyakutake's comet had grabbed the attention of astronomers around the world. Given the brightness at discovery it was possible that this object could reach first magnitude at close approach.

The Orbit

The elements of Hyakutake's orbit are shown in Table 10.1. Shortly after these elements appeared I began to assess how this comet would appear from my location at 52° north. For a change those of us in the northern hemisphere would have the best view since the comet's orbit and the position of the Earth were ideal. When Hyakutake first spotted his comet it was well below the ecliptic plane (Figure 10.1). By 1996 March 12 the comet would pass through the ecliptic heading north and it would be 1.3 AU from the Sun. Thirteen days later, on 25 March, it would make the close Earth fly-by at a distance of only 0.102 AU and it would then appear near the zenith for northern observers. Following this close approach the comet would move further

Table 10.1. Orbital elements for comet C/1996 B2 Hyakutake

Epoch: 1996 April 27.0 (JDT = 2450200.5)		
T = 1996 May 1.39500 TT	ω = 130°.17725	J2000.0
e = 0.9997576	Ω = 188°.04538	J2000.0
q = 0.2302210 AU	i = 124°.92332	J2000.0
A_1 = +2.76, A_2 = +0.0716		

north reaching its maximum distance above the ecliptic plane on 22 April, all the time being well seen by northern hemisphere observers. As the comet started back towards the ecliptic plane the elongation would rapidly decrease and northern observers would lose it in the evening twilight sometime around the end of April. The comet would then continue on to perihelion on 1 May at a solar distance of 0.23 AU. The southbound ecliptic crossing (descending node) would occur four days later and it would then be up to southern observers to recover the comet as it rose out of the morning twilight in the second week of May.

While the orbital circumstances were good there were more serious problems much nearer home. For much of the northern hemisphere the months of February and March are not particularly favourable for comet observers since cloud cover is a regular nuisance. We knew from the orbit that the best of the display would be in the short period between 22–27 March and it was entirely possible that we would

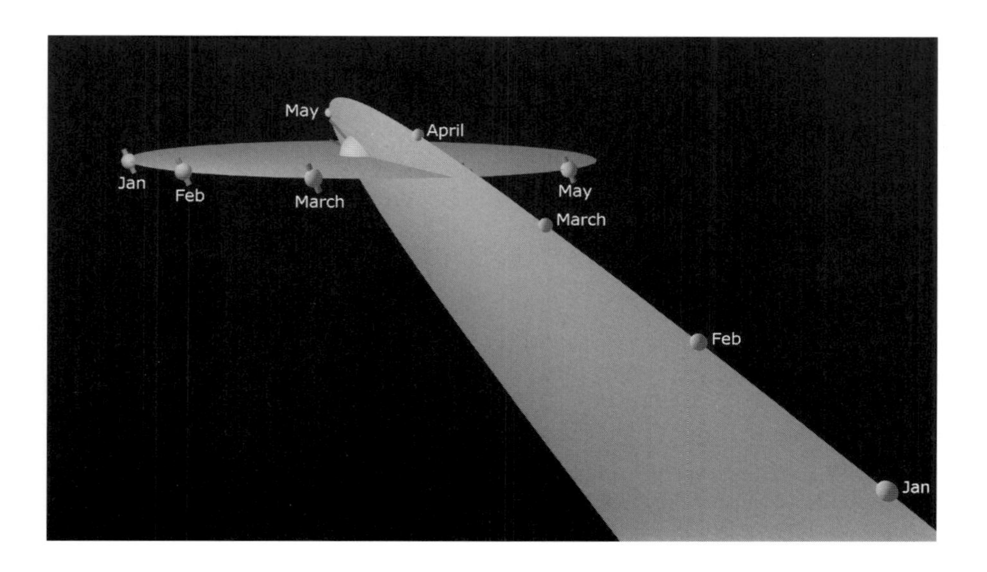

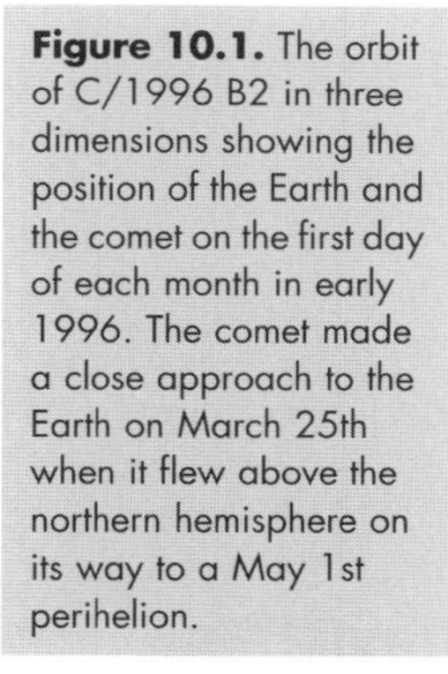

Figure 10.1. The orbit of C/1996 B2 in three dimensions showing the position of the Earth and the comet on the first day of each month in early 1996. The comet made a close approach to the Earth on March 25th when it flew above the northern hemisphere on its way to a May 1st perihelion.

be clouded out completely at the critical time. It was time to start making plans.

Observations

From the viewpoint of mid-northern observers the comet was initially low in the morning sky. Observers further south fared better and many images began to appear on the Internet. By 8 February the 1.5 m ESO telescope at La Silla, Chile, had obtained a spectrum of the comet. This spectrum was dominated by reflected solar radiation but the expected cometary emission from cyanogen (CN) and carbon molecules (C_2 and C_3) was present. Observers at Lowell Observatory reported that the comet's water production rate was around 70% of comet Halley at an equivalent solar distance. This was encouraging news since it implied an active nucleus.

By the second week of February 1996, the magnitude had increased to around 9 and the coma was around 6 arc minutes in diameter. The comet was still a southern hemisphere object at declination –24° and so it remained a difficult object for observers in northern Europe. All the same, the comet was showing encouraging signs of activity and a tail was first detected in a CCD image obtained on 16 February using the Danish 1.54-m telescope at La Silla (Figure 10.2).

The brightness of the comet was increasing rapidly and by 20 February it had reached seventh magnitude and the coma diameter was around 22 arcminutes with a 1° tail. For northern hemisphere observers the comet was south of declination –22°. By early March the comet was still well south of the equator but it had become a naked-eye object at fifth magnitude and binocular observers reported a short tail.

The rotation of the nucleus was inferred from images obtained at Pic-du-Midi on 9 March. These images, taken with a 1.05 m telescope, showed at least two curved jets rotating clockwise and changing orientation over periods of a few minutes. The observed jets were around 2,000 km long and the observations implied that the nucleus had a rotation period of around 6.6 hours.

Radar contact with the comet was achieved on 24 March. Powerful radio pulses were directed towards the comet from the Goldstone dish and echoes were received 107 seconds later. The radar results implied a very small nucleus of 1–3 km diameter surrounded by a

Figure 10.2. The first image to show an ion tail in C/1996 B2 Hyakutake. This is a 15 minute CCD exposure obtained by ESO astronomer Ferdinando Patat on 1996 February 16 using the Danish 1.54 m telescope and a 2K × 2K CCD at La Silla. The field of view is 10.4 arcminutes square with north up. Courtesy European Southern Observatory.

dense cloud of pebble-sized objects. Shortly afterwards the first ever detection of X-rays from any comet was made using the ROSAT satellite. The brightest parts of the comet in X-rays were diffuse, crescent shaped and offset sunwards by about 6 arcminutes from the nucleus corresponding to a distance of about 30,000 km.

By 20 March, when the comet passed into the northern hemisphere, it had brightened to second magnitude and was a beautiful naked eye sight to those lucky enough to have clear skies. The coma was already larger than 1° and an ion tail of up to 20° had been reported by some observers. The comet was now moving quickly across the sky as it approached the Earth. By this time I had decided that my only chance to see this comet was to flee the UK in search of better weather. In this regard Tenerife is a convenient location for UK observers since flights are cheap and plentiful, the holiday infrastructure is well developed, and the weather prospects and latitude were ideal for the comet. A group of us combined our resources and booked the tickets.

Having decided to travel to a better observing site I had to decide what equipment to take. At the time of closest approach Hyakutake was going to be huge so simple SLR cameras and a driven mounting, such as those described in Chapter 5, were all that was required. We took a combination of cameras, a home-made barndoor mount and a larger Vixen SP mounting for longer focal length shots along with lots of film, both colour and hypered Tech Pan. Airport X-ray machines can be quite a problem for travelling astrophotographers since high speed films can be affected if they are scanned. I would always suggest carrying rolls of film in your pocket so that they don't pass through the machines. If you don't have enough pockets you can buy special shielding bags which will protect the film as it is scanned.

For observers at mid-northern latitudes the comet was at its best in the early morning hours as it rose towards the zenith. From Tenerife, on the early morning of 23 March the comet had the distinctive "spring-onion" appearance shown in Figure 10.3. The star Arcturus was embedded in a tail which extended over at least 20°. In 11 × 80 binoculars the coma showed a stellar nucleus surrounded by a classic hood. That evening the tail was less well defined but the coma was slightly brighter than the previous day.

At its closest on the morning of 25 March the comet approached zero magnitude and the tail stretched from Ursa Major, through Boötes and possibly as far as the bowl of Virgo. Fortuitously, the comet become most active right at the time of closest approach and the tail detail that we could see on the night of 24/25 March was astounding. This time we were at an altitude of 1,600 m on Mount Teide and the sky was perfect from dusk until dawn. The visible tail extended for at least 25° with direct vision and well over 40° with averted vision (see Figure 5.9 on page 106). In 11 × 80 binoculars a bright, tailward pointing spike was very prominent and a major disconnection event was clearly visible to the naked eye. On that night the comet totally dominated the sky and it was easy to understand how ancient people must have been terrified by such objects.

The exact time of close approach was 7^h UT on 25 March. After this the comet began to recede from the Earth as it headed in towards the Sun. The viewing geometry meant that the apparent tail-length was expected to grow as the comet moved away from the Earth. In the early morning hours of 26 March we made our final trip to observe the comet from Tenerife. The tail had again changed considerably from the day before and it was longer, possibly up to 50° with averted vision. The comet was now near to Kochab in Ursa Minor.

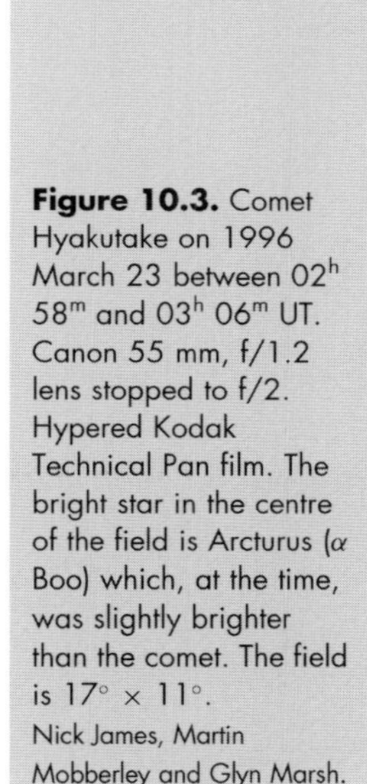

Figure 10.3. Comet Hyakutake on 1996 March 23 between 02^h 58^m and 03^h 06^m UT. Canon 55 mm, f/1.2 lens stopped to f/2. Hypered Kodak Technical Pan film. The bright star in the centre of the field is Arcturus (α Boo) which, at the time, was slightly brighter than the comet. The field is $17° \times 11°$.
Nick James, Martin Mobberley and Glyn Marsh.

On the night of 26/27 March the head passed within 4° of Polaris and various observers took advantage of this to produce some good fixed-camera photos (Figure 5.4 on page 95). By 29 March the Moon had started to become a problem for visual observers but the comet was bright enough to allow relatively simple equipment to capture a spectrum. One such was obtained by Maurice Gavin on 1 April (Figure 10.4). The coma shows reflected sunlight and two prominent Swan Band emission lines in the cyan and green parts of the spectrum. Two fainter emission lines are visible in the blue and yellow and absorption lines due to Earth's atmosphere are visible in the red. Since the tail was much fainter only the reflected solar continuum is visible.

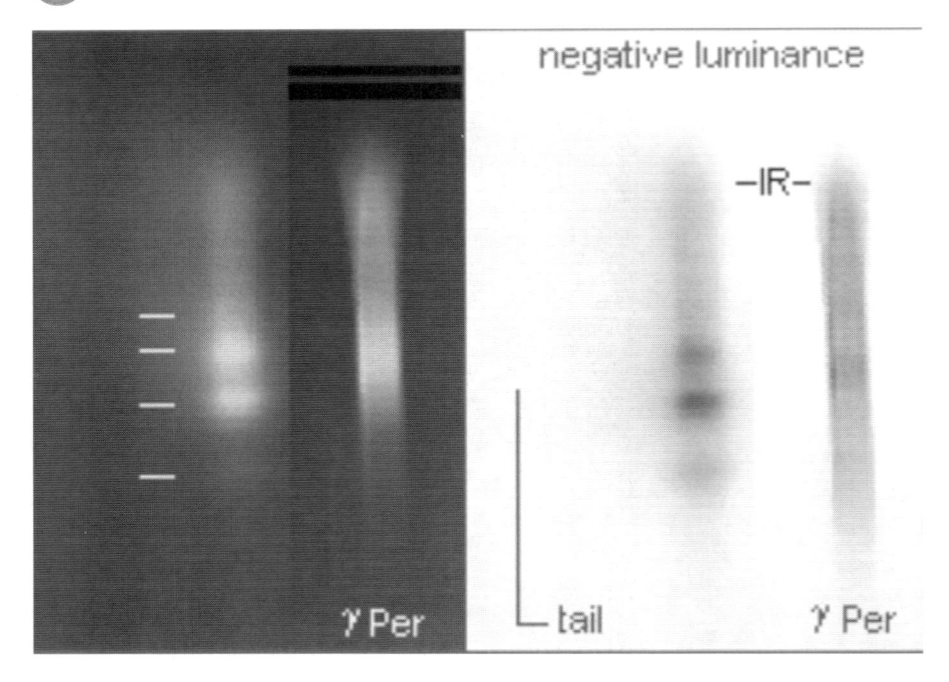

Figure 10.4. Low resolution spectrum of C/1996 B2 obtained on 1996 April 1 using an 85 mm f/2 lens with an objective prism and an SX colour CCD. This spectrum shows the prominent Carbon molecule emission lines. Maurice Gavin.

By the end of the first week of April 1996, the comet was sinking lower into the north-west on its way to perihelion. A prominent dust tail finally began to appear around 10 April although it never surpassed the ion tail. Meanwhile jet activity in the coma continued. The tail now had a classic appearance with a sharp ion tail and a diffuse dust tail. Photographs taken with Schmidt cameras reveal exquisite detail and many streamers are visible in the tail (Figure 10.5).

As the comet moved towards the Sun it entered the field of the LASCO C3 coronograph on board the SOHO satellite. Images of the comet at perihelion were obtained by this instrument (Figure 10.6) but terrestrial observers had to wait until 9 May for the comet to reappear from the Sun's glare. By this time it was a southern hemisphere morning object at around third magnitude. The comet faded rapidly as it moved away from the Sun. Southern hemisphere observers picked it up on 9 May in a bright sky. By 18 May the comet was visible in a darker sky but the coma had faded to fourth magnitude. It faded past sixth magnitude around 10 June and by late August it was fainter than magnitude 10.

Since Hyakutake made such a close approach to the Earth it was possible to see considerable detail in the inner coma with amateur-sized telescopes. The comet

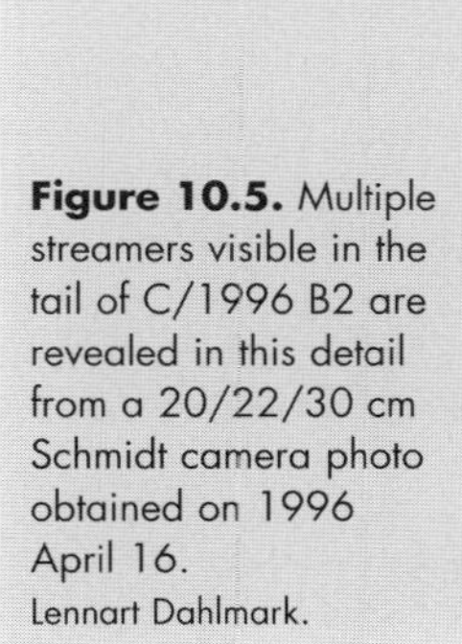

Figure 10.5. Multiple streamers visible in the tail of C/1996 B2 are revealed in this detail from a 20/22/30 cm Schmidt camera photo obtained on 1996 April 16.
Lennart Dahlmark.

showed many interesting features close to the nucleus. Of particular importance were a sunward fan and the very intense tailward-pointing jet that was seen around close approach. This was also the first bright comet to make a close approach since CCD cameras became widely available in the amateur community. Photographers had never been particularly successful in reproducing the details visible in the inner coma of comets since the wide variation in light levels exceeded the available dynamic range of printing papers. The ability of CCD cameras to operate over a very large range of brightness levels meant that they could record the fine detail in the inner coma which in former times was only available to visual observers. Advanced processing techniques such as unsharp masks and radial filters can be applied easily to electronic images (as described in Chapter 8) and some of the amateur CCD results were stunning.

Terry Platt obtained a sequence of CCD images on 1 April using an SX camera. The exposures were kept

Figure 10.6. Comet C/1996 B2 near perihelion captured by the LASCO C3 coronagraph on board SOHO. Image obtained on 1996 May 3. Courtesy of SOHO/LASCO consortium. SOHO is a project of international cooperation between ESA and NASA.

short to ensure that the CCD did not saturate on the bright inner coma and each frame was processed with an unsharp mask. The nine frames clearly show the anti-clockwise rotation of various features (Figure 10.7) and the effect is particularly strong when the frames are processed into a movie (see the CD-ROM).

One of the most impressive features of this comet was its tail. Since Hyakutake made such a close approach its tail aspect changed rapidly from the end of March to mid-April. It was also possible to observe major changes in the tail structure over periods measured in hours. We do not normally have the opportunity to do this since comets with large tails are usually only visible in a dark sky for a short period. Hyakutake had the unusual property that, at the time of close approach, its massive tail was visible for the entire night.

Hyakutake was certainly a Great Comet but its discoverer was a particularly modest man. In a statement released shortly after the discovery he said: "I am a bit perplexed by all the attention paid to me, when it is the comet that deserves the credit."

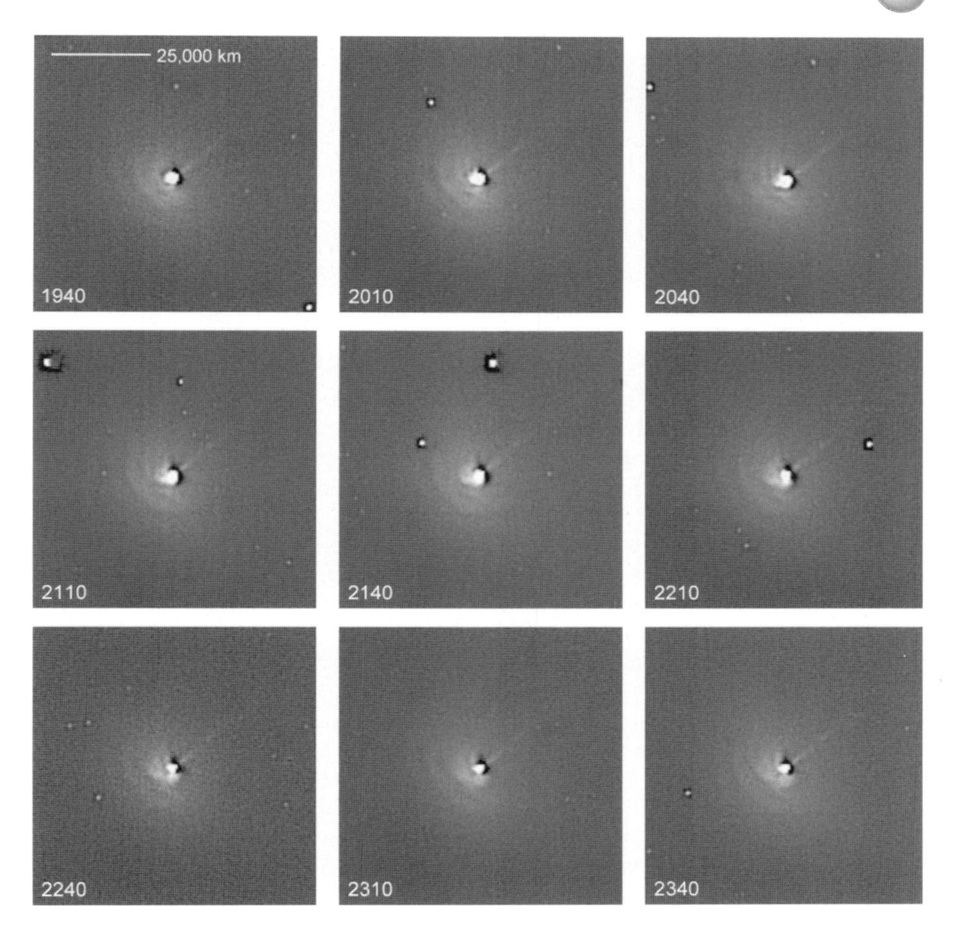

Figure 10.7. Images of the inner coma of C/1996 B2 Hyakutake taken on 1996 April 1 show a clear anti-clockwise rotation over a period of four hours. Changes in the tailward pointing jet are also visible. Terry Platt.

The Discovery of Hale–Bopp

On the night of 1995 July 22/23, around six months before Hyakutake's second discovery, Alan Hale and Thomas Bopp made independent discoveries of the comet which would become one of the most famous astronomical objects in history. At the time the comet was near to M70 in the star-clouds of Sagittarius and it was visible as a faint tenth magnitude fuzz. It was quickly designated C/1995 O1 (Hale–Bopp) but it was a while before a good orbit was available. At discovery

Table 10.2. Orbital elements for C/1995 O1 Hale–Bopp

Epoch: 1997 March 13.0 (JDT = 2450520.5)		
T = 1997 April 1.1373 TT	ω = 130°.5887	J2000.0
e = 0. 995068	Ω = 282°.4707	J2000.0
q = 0. 914142 AU	i = 89°.4300	J2000.0
A_1 = +1.27, A_2 = +0.1144		

Hale–Bopp was moving very slowly to the NW and one possibility was that the comet was heading straight for us. This led to "Doomsday Comet" headlines in the newspapers but when the preliminary orbital elements were announced on 27 July they showed that, far from being a close-by object moving towards us, Hale–Bopp was a far-away monster that would not approach our planet any closer than 1.2 AU.

The orbital elements are listed in Table 10.2. At the time of its discovery Hale–Bopp was over 7 AU from the Sun and 6.2 AU from the Earth. If it was tenth magnitude at that distance how bright would it be at perihelion? Now that he had a preliminary orbit Rob McNaught at the Anglo-Australian Observatory could search old Schmidt plates for prediscovery images. The oldest image of the comet the he could find was on a plate taken on 1993 April 27 using the UK Schmidt. At that time the comet was 13.1 AU from the Sun at a magnitude of around 18. There was now no doubt that this was a huge comet. McNaught's old positions helped to improve the orbit and they showed that Hale–Bopp had last visited the inner Solar System around 4,200 years ago. Perturbations this apparition would change the orbit so that we would have to wait only 3,400 years until its next visit.

The circumstances of Hale–Bopp's apparition were especially favourable for those of us in the northern hemisphere. When it was discovered Hale–Bopp was further away than Jupiter and well south of the ecliptic (Figure 10.8) at a declination of –32°. The plane of the orbit is almost at right angles to the ecliptic and the comet would move slowly towards its northbound crossing (the ascending node). It reached this on 1996 February 29 by which time it was 5.2 AU away from the Sun and 5.8 AU from the Earth but still at a declination of –22°. At around this time, in January 1996, it went behind the Sun as seen from the Earth (conjunction) but it was recovered in the morning sky in early February. The next opposition occurred in 1996 June

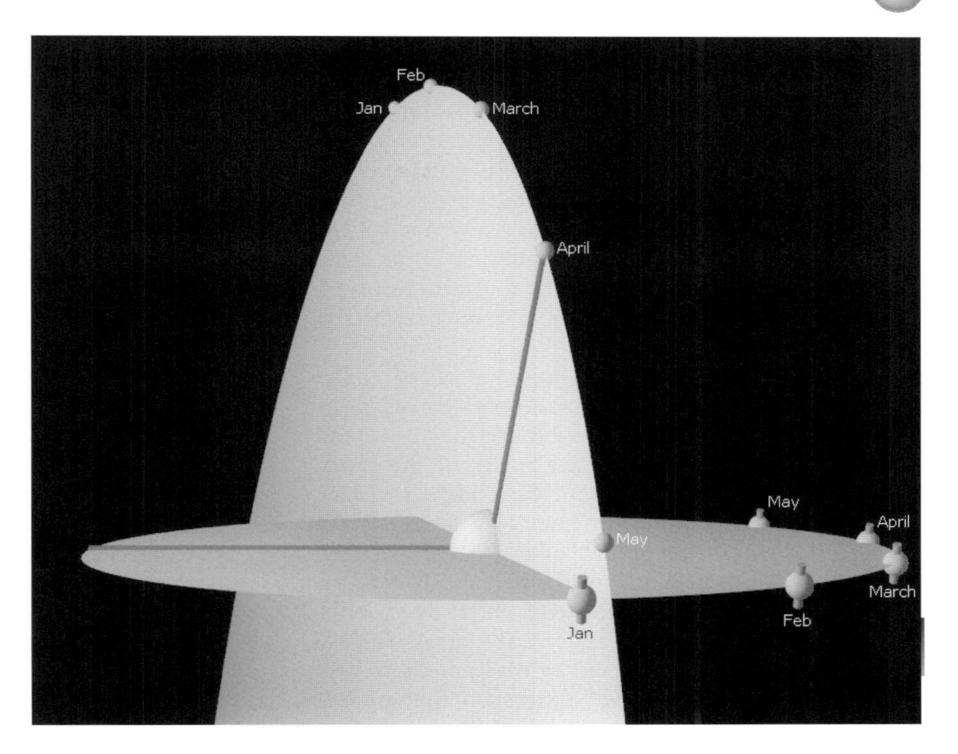

Figure 10.8. The orbit of C/1995 O1 in three dimensions showing the position of the Earth and the comet on the first day of each month in early 1997. The comet was discovered when it was well south of the ecliptic but it was at its brightest when it was north of the ecliptic plane. Perihelion occurred on April 1st.

when the comet was in Scutum at a declination of –20° and the distance to the Sun and Earth was 3.9 and 2.9 AU respectively. During the autumn and winter of 1996 the comet gradually became brighter and it moved north passing over the celestial equator in early December. This was about the time of the next conjunction but the elongation never became less than 26° and so observations were able to continue without a break as the comet was followed into the morning sky.

In early 1997 the comet moved rapidly north and it reached the furthest distance above the ecliptic on 6 March when it was in Cygnus. By this time the comet was best seen as an evening object. At perihelion on 1 April Hale–Bopp was 0.91 AU from the Sun, 1.35 AU from the Earth and at a declination of +44°. The comet was then moving rapidly south and northern hemisphere observers lost it at the end of May. Hale–Bopp crossed the ecliptic at the descending node on 1997 May 7 but by this time the Earth was almost on the opposite side of the Sun. The comet crossed back into the southern celestial hemisphere at the end of June and it reached a declination of –64° by the end of the 1997.

Observations

It was clear that Hale–Bopp was going to be impressive. Throughout the latter part of 1995 and the early part of 1996 the comet steadily brightened. Professionals were busy observing this new object and the first detection of CO took place in September 1995 when it was 6.7 AU from the Sun. CO is the dominant molecule in the coma this far from the Sun since water only starts to play a significant part in the activity of a comet at around 3 AU.

By the summer of 1996 Hale–Bopp was visible to the naked eye. From the time of its discovery a considerable amount of detail had been visible in the coma and observers became perplexed that many of the radial features appeared to be constant from night to night. Some even suggested that the nucleus rotation period was greater than a year! Much of this detail was seen visually but it was also well recorded on CCD images. The bizarre appearance was explained by Zdenek Sekanina of NASA's Jet Propulsion Laboratory. He said that the jets could be explained by dust emission from three or four discrete active sources on the rotating nucleus which were periodically switched on and off as they rotated in to and out of the sunlit side of the comet (Figure 10.9).

From careful analysis of the jets it seemed that the nucleus had a rotation period of around 11.5 hours. This was confirmed by observation of the concentric dust shells in the inner coma which originated from the active jets on the nucleus. These were separated by around 12,000 km and they were expanding outwards at 0.3 km/s which was consistent with the 11.5 hour spin period.

The last solar conjunction prior to perihelion occurred around Christmas 1996 but by this time Hale–Bopp was at such a high northern declination that many observers followed it all the way through the winter. In 1997 January the show really started. Hale–Bopp began to dominate the morning sky with its growing tails and extraordinary coma detail. By March it was moving into the evening sky as an unmissable object visible to anyone who bothered to look. It had a total magnitude of around –0.6 and its bright dust tail was even visible from light-polluted cities. From the

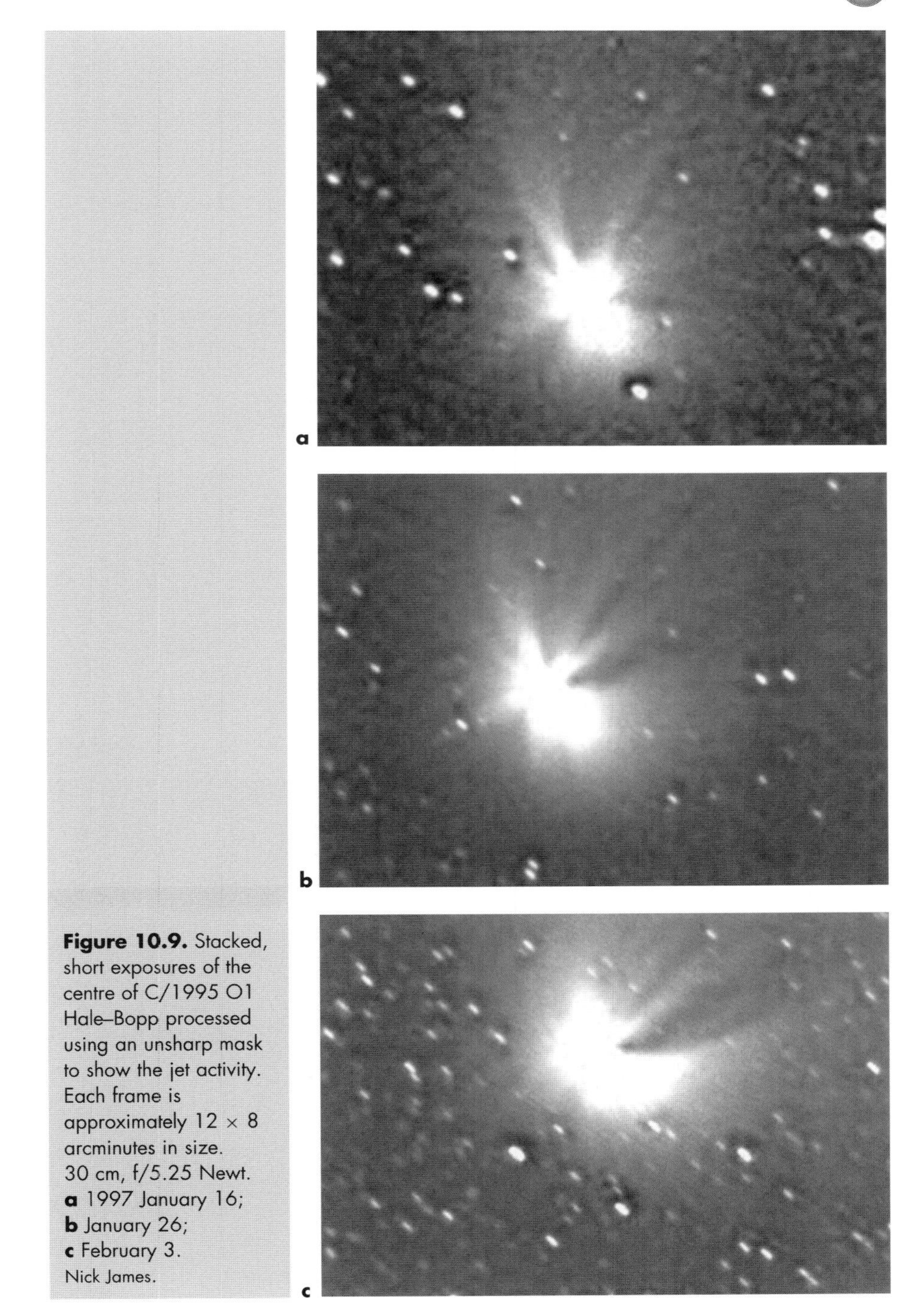

Figure 10.9. Stacked, short exposures of the centre of C/1995 O1 Hale–Bopp processed using an unsharp mask to show the jet activity. Each frame is approximately 12 × 8 arcminutes in size. 30 cm, f/5.25 Newt. **a** 1997 January 16; **b** January 26; **c** February 3.
Nick James.

countryside the entire comet including its long, straight ion tail was a stunning sight. The pictures on these pages demonstrate that the comet was an amazing object at whatever scale we looked. It had fantastic features that were visible in the centre of the coma with a large telescope (Figure 10.10) but it also had a stunning beauty that was only visible in wide-angle views (Figure 10.11).

A very interesting observation was made in early April 1997 when astronomers using the James Clerk Maxwell telescope on Mauna Kea, Hawaii detected a special form of water, HDO, in Hale–Bopp. Most cometary activity comes from normal water, H_2O, but in some cases the one of the Hydrogen atoms in the water molecule gets replaced by an isotope called Deuterium. The ratio of Deuterium to Hydrogen is a very powerful indicator of where the water came from rather in the manner of a DNA fingerprint. The team found that the ratio in Hale–Bopp was one part of Deuterium to 10,000 parts of Hydrogen which is comparable to ratio found in the Earth's oceans but different to the ratio seen in the interstellar medium. This lends support to the theory that all of the water on our planet originally came from impacting comets.

Figure 10.10. A sequence showing almost one full rotation of the nucleus of C/1995 O1 Hale–Bopp on 1997 March 28/29. The images were obtained between 19^h 00^m UT on the 28th and 04^h 50^m UT on the 29th using a one-shot Starlight Xpress colour camera. Each image is approximately 2 arcminutes square. Terry Platt.

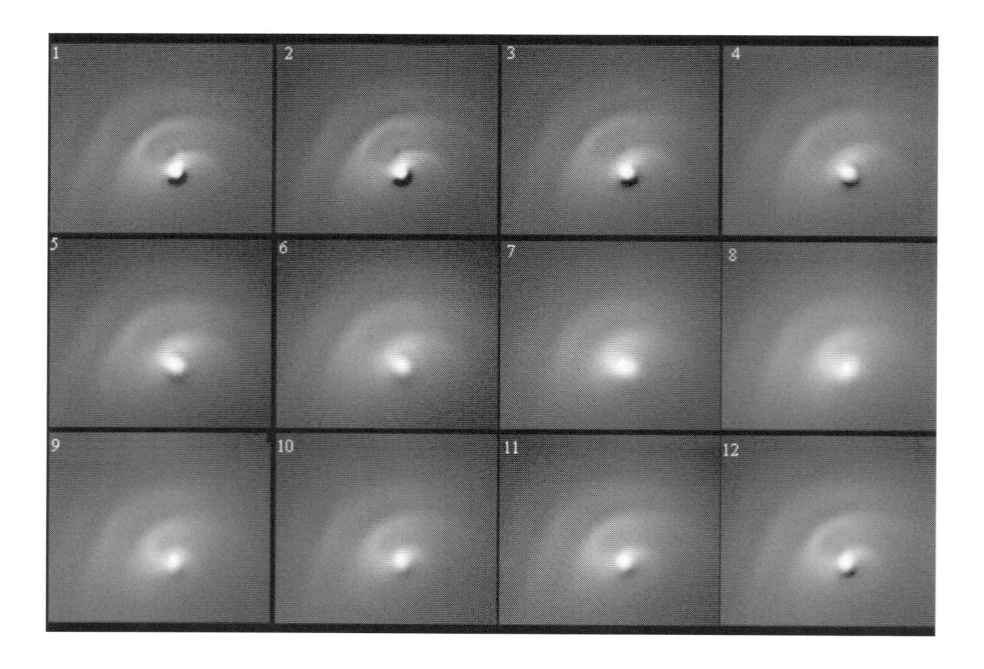

Figure 10.11. A wide field view of Hale–Bopp taken on 1997 March 16 at 03^h 11^m UT. 25 minute exposure on 120 format Fuji SG+ using an 80 mm FL lens. Mauro Zanotta.

The Tails

Hale–Bopp had no fewer than three tails. In addition to the expected type I (ion) and type II (dust) tails a team using the Isaac Newton Telescope on La Palma detected a large tail of neutral sodium atoms. This sodium tail was seen as a straight feature 6°.6 long and less than 10 arcminutes wide with parallel edges over its entire length. No such tail had ever been seen in previous comets. Spectra showed that the tail was pure sodium with none of the normal cometary components such as water or carbon present. Theory showed that the sodium could not have been released directly from the nucleus but it must have arisen from other molecules which were dissociated after ejection.

The dust tail in most comets is normally rather featureless but, as mentioned in Chapter 5, some large comets display interesting features called *synchronic bands.* These bands are thought to be caused by episodic releases of dust from the comet's nucleus as sources turn on and off when they rotate in to and out of sunlight. The dust particles released are then affected by solar radiation pressure in different ways depending on their size. Smaller particles get accelerated by more than larger ones and a single release of dust gets spread out along a line called a *synchrone* to form a well-defined linear feature called a *synchronic band*. Dust which is released continuously from the nucleus moves along an arc called a *syndyne* and this defines the normally rather bland and featureless background of the dust tail.

Only a few comets have exhibited synchronic banding. Prior to Hale–Bopp synchronic bands had been seen most prominently in the glorious dust tail of Comet West in March 1976. The bands in that comet were especially prominent because the nucleus had broken up. No such damage afflicted Hale–Bopp's nucleus but since the comet was so large and active there were plenty of episodic dust releases to generate the telltale bright diagonal striations of synchronic banding.

During the early part of 1997 the position of the comet relative to the Earth and Sun meant that Hale–Bopp's dust tail was pointing away from us and so very little detail was to be seen. As the geometry became more favourable during the early spring synchronic bands became visible in the dust tail as it lengthened and broadened. Glyn Marsh and Denis Buczynski managed to image these features using various astrographic lenses of 0.5 to 1.2 metre focal length and large format (5-inch × 4-inch and 10-inch × 8-inch) high resolution sheets of Hydrogen hypered Kodak Technical Pan film (Figure 10.12).

Professional workers at Calar Alto also detected at least 12 straight bands in the tail on 17/18 March ranging from 1°.5 to 3° from the nucleus. These bands were 1 to 3 arcminutes wide and some of them extended more than 1° across the tail. The position angle of these bands indicated that they were not in exactly the place expected for a synchrone assuming direct emission from the nucleus and so the dust grains had probably come from secondary processes in the outer coma or the dust-tail.

On its Way Out

By August 1997 Hale–Bopp was only visible from the southern hemisphere and it had faded to third

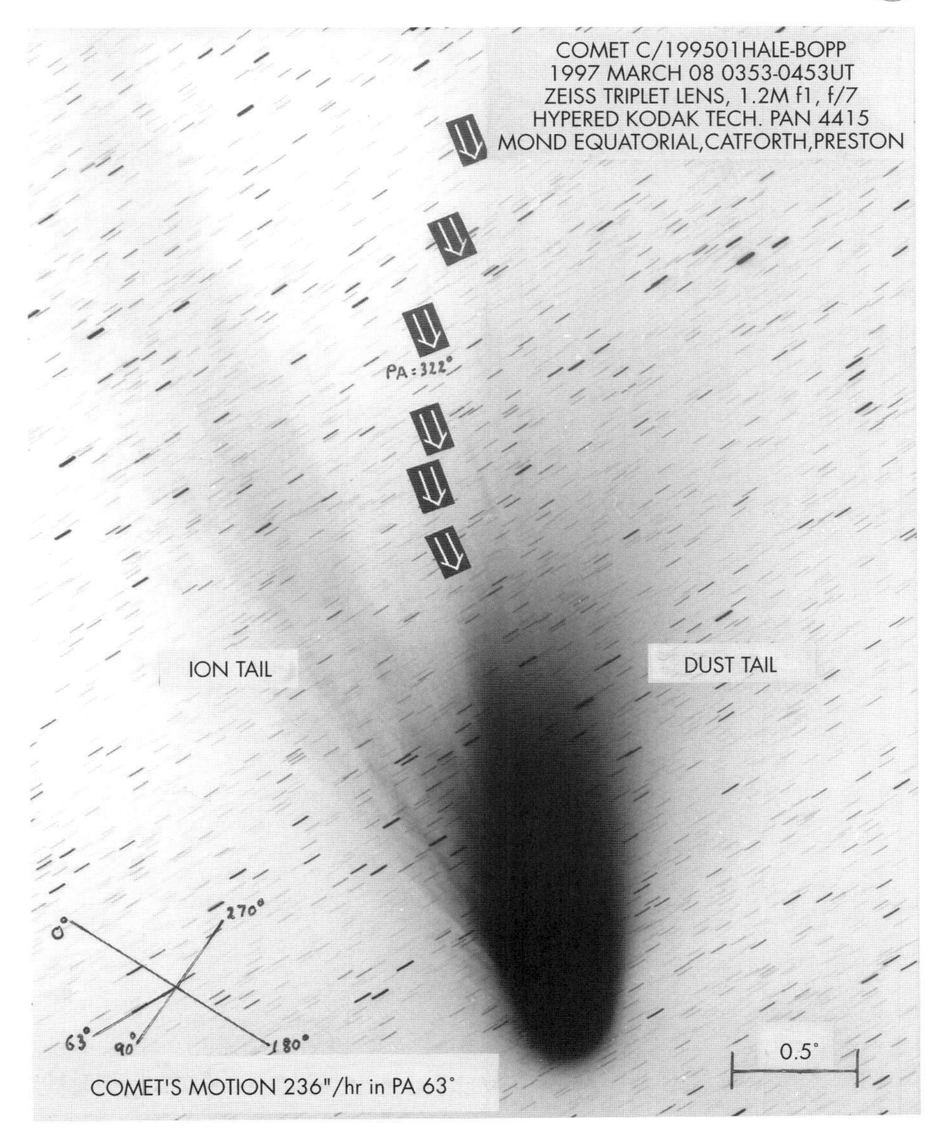

Figure 10.12. Comet C/1995 O1 Hale–Bopp showing synchronic bands in the dust tail. 1997 March 8 03^h 53^m to 04^h 53^m UT. 1.2m FL, f/7 Zeiss Triplet lens mounted on the Mond Equatorial. Hypered Kodak Tech Pan 4415 film.
Glyn Marsh and Denis Buczynski.

magnitude. By the end of 1997 it had fallen back below naked-eye visibility but it did develop a short anti-tail for a while as it crossed the ecliptic.

As the comet faded most people stopped observing it and this is all the encouragement that a comet needs to put on an unexpected show. By 1998 December the comet was more than 7 AU from the Sun and the total magnitude was around 11. Gordon Garradd was still monitoring it from Australia and he reported a remarkable 3-magnitude brightening of the nuclear region. The total magnitude of the comet was not much affected but this nuclear outburst gradually enlarged and dispersed through the coma. This was very similar to the outbursts seen in comet 29P/Schwassmann–Wachmann 1 at comparable solar distances of 6–7 AU and it shows how important it is to continue monitoring comets for as long as possible as they retreat from the Sun.

Hale–Bopp remains visible as it travels back to the outer Solar System (Figure 10.13). At the end of 2001 it was 15 AU from the Sun but still had a total magnitude of 15. It was then in the far south constellation of Mensa but large telescopes were still tracking it. It will reach 30 AU out at the end of 2010 when it will only be a few degrees from the southern pole. Even then the comet should still be visible to CCD equipped amateur telescopes at a predicted predicted magnitude of 19. A

Figure 10.13. Comet C/1995 O1 Hale–Bopp on 1999 June 18 with the EMMI instrument attached to the ESO 3.5 m New Technology Telescope (NTT) at La Silla. At this time the comet was about 8.7AU from both the Earth and the Sun. The field of view is about 7.2 arcminutes square with north up. Courtesy European Southern Observatory.

full twenty years after its discovery it should fall below magnitude 20 but it will remain visible in large telescopes for decades to come.

What's Next?

Astronomers have learned a great deal about comets and how they fit into the grand scheme of the Solar System. Of course, there is still much more to learn. What will the future bring?

Maybe one day men will walk on a cometary nucleus but this would be a risky thing to do if they chose a time when the comet was active. Their spacecraft would be subject to a hazardous shot-blasting as it entered the coma and the nucleus itself would provide a very treacherous landing pad. While manned landings may happen in the far-distant future our unmanned robots will be making contact within the next few years.

Perhaps the most exiting project is the European Space Agency's Rosetta probe (Figure 10.14). This will be launched in January 2003 and the probe will rendezvous with comet 46P/Wirtanen in November 2011 after a complex sequence of planetary fly-bys. Rosetta will enter a mapping orbit and should get as close as one kilometre to the comet's nucleus. In 2012 a small lander will be deployed to settle on the icy nucleus and make the first in-situ measurements from the surface of a comet.

While Rosetta will be the first soft-lander it will not be the first item of human hardware to make contact with a comet. The NASA Deep Impact mission will be launched after Rosetta but it will rendezvous with comet 9P/Tempel 1 in July 2005. The probe will approach the comet at a closing speed of over 10 km/s and it will release a 350 kg impactor which will strike the nucleus and form a crater possibly 100 m in diameter and 25 m deep. The impact will be observed by the probe and the on-board equipment should give us vital information about the composition of the nucleus.

One probe that is already on its way is NASA's Stardust. This will approach comet 81P/Wild 2 in early 2004 so that it can collect some dust from the comet's tail in a sticky substance called aerogel. The dust, inside a sealed capsule, will be returned to the Earth in 2006. Scientists will be able to analyse this material in great detail and it should help to provide answers to some of

Figure 10.14. An artist's impression of the *Rosetta* orbiter and lander approaching comet 46P/Wirtanen in late 2011.
Copyright ESA.

the biggest questions we have about comets. For instance some scientists hold that the life could never have originated by itself on our planet; they say that the conditions on the early Earth were just too extreme for the fragile components of life to have survived. Many scientists now think that the water in our oceans was delivered to the Earth by cometary impacts. Could the primordial building blocks of life also have arrived the same way?

There is still plenty to learn about the underlying physics and chemistry, and in particular the dynamics, of the comets themselves. Most of that research can be carried out from the surface of the Earth. Add to that the fact that each comet is unique and you can see why the astronomy of comets is so fascinating. While many of the advances in our understanding will be made by professional astronomers the amateur still has a major role to play. It is certainly possible for you to undertake scientifically useful work and we hope that this book

will help you in that endeavour. Most of all, though, we hope that you will enjoy the thrill of seeking out and observing those eerie transient visitors that arrive from the depths of space.

Appendix

Here is a limited list of resources to help you pursue your study of comets and the techniques and equipment necessary to observe them. By making use of your local library, browsing publishers catalogues and the adverts for equipment suppliers, perusing astronomical magazines and periodicals and, above all, by surfing the Internet while making use of a search engine such as *Google.com*, you can uncover plenty more!

Books

Advanced Amateur Astronomy, Gerald North, Cambridge University Press, 1997.

Astronomical Equipment for Amateurs, Martin Mobberley, Springer-Verlag, 1999.

CCD Astronomy: Construction and use of a CCD camera, Christian Buil, Willmann-Bell, 1991.

The Art and Science of CCD Astronomy, edited by David Ratledge, Springer-Verlag, 1997.

Introduction to Comets, J.C. Brandt and Robert C. Chapman, Cambridge University Press, 1983.

Comets, Meteors and Asteroids, John Man, BBC Consumer Publishing, 2001.

Comet Science, Therese Encrenaz, Cambridge University Press, 2000.

Great Comets, Robert Burnham, Cambridge University Press, 2000.

Comets: Creators and Destroyers, David Levy, Touchstone Books, 1998.

The Comet Hale–Bopp, Robert Burnham, Cambridge University Press, 1997.

Publications of astronomical organizations

(contact web addresses given further on)

ICQ Guide to Observing Comets, edited by Dan Green.
BAA Comet Section Observing Guide, edited by Jonathan Shanklin.

Software

Image processing programs

Astroart	http://www.msb-astroart.com/down_en.htm
Mira	http://www.axres.com/
AIPWIN	http://www.willbell.com/aip4win/AIP.htm
MaxIm DL	http://www.cyanogen.com
IRIS	http://astrosurf.com/buil/us/iris/iris.htm
Starlink	http://star-www.rl.ac.uk/
IRAF	http://iraf.noao.edu/iraf-homepage.html

Observation planning

Guide	http://www.projectpluto.com
Grafdark	http://www.naas.btinternet.co.uk/grafdark.htm
JPL horizons	http://ssd.jpl.nasa.gov/horizons.html
Heavens above	http://www.heavens-above.com
Observable comets	http://cfa-www.harvard.edu/iau/Ephemerides/Comets/index.html
The Sky	http://www.bisque.com/Products/TheSky/TheSky.asp

Orbit computation and integration

Mercury	http://star.arm.ac.uk/~jec/home.html
Orbfit	http://newton.dm.unipi.it/orbfit
N-body integrators	http://www.scitec.auckland.ac.nz/~sharp/n-body_integrators/

Miscellaneous

MPEG encoders	http://bmrc.berkeley.edu/frame/research/mpeg/mpeg_encode.html
	http://www.mpeg.org

NTP time clients	http://www.astrosurf.com/astropc/timememo/ http://www.ntp.org

CCD cameras

Commercial cameras

Starlight XPress	http://www.starlight-xpress.co.uk/
SBIG	http://www.sbig.com
Apogee	http://www.apogee-ccd.com/

Do-it-yourself cameras

Audine	http://astrosurf.com/audine/
Cookbook	http://www.wvi.com/~rberry/cookbook.htm

Spectroscopy

Buil	http://astrosurf.com/buil/us/spe1/spectro1.htm
Spectr'aude	http://astrosurf.com/buil/us/spectro8/spaude_us.htm
Maurice Gavin	http://www.astroman.fsnet.co.uk/spectro.htm

Comet information

Catalog of cometary orbits	http://cfa-www.harvard.edu/iau/services/CCat.html
SOHO discoveries	http://www.ph.u-net.com/comets/comet-links.htm
CBAT computer service	http://cfa-www.harvard.edu/iau/services/CS.html

Astronomical alert services

IAU circulars	http://cfa-www.harvard.edu/iau/services/Subscriptions.html
TA circulars	http://www.theastronomer.org/subscription.html
Sky and Telescope	http://skyandtelescope.com/observing/proamcollab/astroalert

Useful hardware

Micropositioners	http://www.catalog.coherentinc.com

Comet images on the web

Wil Milan	http://www.astrophotographer.com/cometphotos.html
Herman Mikuz	http://www.fiz.uni-lj.si/astro/comets/
Terry Lovejoy	http://users.bigpond.net.au/southern/southern.htm
Gerald Rhemann	http://www.astrostudio.at/defaultIE.htm
JPL comet page	http://encke.jpl.nasa.gov/
The Astronomer	http://www.theastronomer.org/comets.html

Miscellaneous

Ceravolo video	http://www.andromedasoftware.com/cd240.htm
3D free-fusion	http://www.slc.edu/~ebj/sight_mind/stereo/free_fusion.html
Hypered film	http://www.lumicon.com

Astronomical organizations

The Astronomer	http://www.theastronomer.org
BAA comet section	http://www.ast.cam.ac.uk/~jds/
ALPO comet section	http://www.lpl.arizona.edu/~rhill/alpo/comet.html
German comet sect.	http://www.fg-kometen.de/fgk_hpe.htm
ICQ	http://cfa-www.harvard.edu/icq/icq.html

Index

The Accompanying CD-ROM

The accompanying CD-ROM is compatible with a wide range of computers and all of the data on this disk can be accessed using your favourite web browser. The pages are designed to use a very basic form of HTML which should be supported by practically every browser and operating system. The disk itself is mastered using the ISO 9660 level 2 filesystem since this is universally supported by those operating system which support CD-ROM drives. To start simply click on the index.html file in the root directory of your CD-ROM drive.

Each page on the disk contains a contents list in the blue bar to the left of the screen. The links in this contents list lead to various key information sources on the disk. The main text also includes links which will take you to other areas of the disk. Some of the links take you to remote sites on the Internet and so you will need a Net connection for these.

The contents of the disk are as follows:

- Resources – A copy of appendix 1 with clickable links.
- Reporting – How to format and report your observations.
- A collection of CCD and negative scans in FITS format. These can be used to practice your image processing skills.
- A number of programs. These are a program for generating comet ephemerides, a set of programs for performing orbit integration and a command-line image processing utility.
- Finally a comet gallery containing descriptions and images of Hale–Bopp, Hyakutake and many other comets. This gallery also contains some movies and some orbit diagrams showing how comets move in space.